Reviews of Environmental Contamination and Toxicology

VOLUME 160

Springer
New York
Berlin
Heidelberg
Barcelona
Hong Kong
London
Milan
Paris
Singapore
Tokyo

Reviews of Environmental Contamination and Toxicology

Continuation of Residue Reviews

Editor
George W. Ware

Editorial Board
Lilia A. Albert, Xalapa, Veracruz, Mexico
F. Bro-Rasmussen, Lyngby, Denmark · D.G. Crosby, Davis, California, USA
Pim de Voogt, Amsterdam, The Netherlands · H. Frehse, Leverkusen-Bayerwerk, Germany
O. Hutzinger, Bayreuth, Germany · Foster L. Mayer, Gulf Breeze, Florida, USA
N.N. Melnikov, Moscow, Russia · D.P. Morgan, Cedar Rapids, Iowa, USA
Douglas L. Park, Baton Rouge, Louisiana, USA
Annette E. Pipe, Burnaby, British Columbia, Canada
Raymond S.H. Yang, Fort Collins, Colorado, USA

Founding Editor
Francis A. Gunther

VOLUME 160

Springer

Coordinating Board of Editors

DR. GEORGE W. WARE, *Editor*
Reviews of Environmental Contamination and Toxicology

5794 E. Camino del Celador
Tucson, Arizona 85750, USA
(520) 299-3735 (phone and FAX)

DR. HERBERT N. NIGG, *Editor*
Bulletin of Environmental Contamination and Toxicology

University of Florida
700 Experimental Station Road
Lake Alfred, Florida 33850, USA
(941) 956-1151; FAX (941) 956-4631

DR. DANIEL R. DOERGE, *Editor*
Archives of Environmental Contamination and Toxicology

6022 Southwind Drive
N. Little Rock, Arkansas, 72118, USA
(501) 791-3555; FAX (501) 791-2499

Springer-Verlag
New York: 175 Fifth Avenue, New York, NY 10010, USA
Heidelberg: Postfach 10 52 80, 69042 Heidelberg, Germany

Library of Congress Catalog Card Number 62-18595.
Printed in the United States of America.

ISSN 0179-5953

Printed on acid-free paper.

© 1999 by Springer-Verlag New York, Inc.
All rights reserved. This work may not be translated or copied in whole or in part without the written permission of the publisher (Springer-Verlag New York, Inc., 175 Fifth Avenue, New York, NY 10010, USA), except for brief excerpts in connection with reviews or scholarly analysis. Use in connection with any form of information storage and retrieval, electronic adaptation, computer software, or by similar or dissimilar methodology now known or hereafter developed is forbidden. The use of general descriptive names, trade names, trademarks, etc., in this publication, even if the former are not especially identified, is not to be taken as a sign that such names, as understood by the Trade Marks and Merchandise Marks Act, may accordingly be used freely by anyone.

ISBN 0-387-98627-8 Springer-Verlag New York Berlin Heidelberg   SPIN 10692728

Foreword

International concern in scientific, industrial, and governmental communities over traces of xenobiotics in foods and in both abiotic and biotic environments has justified the present triumvirate of specialized publications in this field: comprehensive reviews, rapidly published research papers and progress reports, and archival documentations. These three international publications are integrated and scheduled to provide the coherency essential for nonduplicative and current progress in a field as dynamic and complex as environmental contamination and toxicology. This series is reserved exclusively for the diversified literature on "toxic" chemicals in our food, our feeds, our homes, recreational and working surroundings, our domestic animals, our wildlife and ourselves. Tremendous efforts worldwide have been mobilized to evaluate the nature, presence, magnitude, fate, and toxicology of the chemicals loosed upon the earth. Among the sequelae of this broad new emphasis is an undeniable need for an articulated set of authoritative publications, where one can find the latest important world literature produced by these emerging areas of science together with documentation of pertinent ancillary legislation.

Research directors and legislative or administrative advisers do not have the time to scan the escalating number of technical publications that may contain articles important to current responsibility. Rather, these individuals need the background provided by detailed reviews and the assurance that the latest information is made available to them, all with minimal literature searching. Similarly, the scientist assigned or attracted to a new problem is required to glean all literature pertinent to the task, to publish new developments or important new experimental details quickly, to inform others of findings that might alter their own efforts, and eventually to publish all his/her supporting data and conclusions for archival purposes.

In the fields of environmental contamination and toxicology, the sum of these concerns and responsibilities is decisively addressed by the uniform, encompassing, and timely publication format of the Springer-Verlag (Heidelberg and New York) triumvirate:

Reviews of Environmental Contamination and Toxicology [Vol. 1 through 97 (1962–1986) as Residue Reviews] for detailed review articles concerned with any aspects of chemical contaminants, including pesticides, in the total environment with toxicological considerations and consequences.

Bulletin of Environmental Contamination and Toxicology (Vol. 1 in 1966) for rapid publication of short reports of significant advances and discoveries in the fields of air, soil, water, and food contamination and pollution as well as methodology and other disciplines concerned with the introduction, presence, and effects of toxicants in the total environment.

Archives of Environmental Contamination and Toxicology (Vol.1 in 1973) for important complete articles emphasizing and describing original experimental or theoretical research work pertaining to the scientific aspects of chemical contaminants in the environment.

Manuscripts for *Reviews* and the *Archives* are in identical formats and are peer reviewed by scientists in the field for adequacy and value; manuscripts for the *Bulletin* are also reviewed, but are published by photo-offset from camera-ready copy to provide the latest results with minimum delay. The individual editors of these three publications comprise the joint Coordinating Board of Editors with referral within the Board of manuscripts submitted to one publication but deemed by major emphasis or length more suitable for one of the others.

Coordinating Board of Editors

Preface

Thanks to our news media, today's lay person may be familiar with such environmental topics as ozone depletion, global warming, greenhouse effect, nuclear and toxic waste disposal, massive marine oil spills, acid rain resulting from atmospheric SO_2 and NO_x, contamination of the marine commons, deforestation, radioactive leaks from nuclear power generators, free chlorine and CFC (chlorofluorocarbon) effects on the ozone layer, mad cow disease, pesticide residues in foods, green chemistry or green technology, volatile organic compounds (VOCs), hormone- or endocrine-disrupting chemicals, declining sperm counts, and immune system suppression by pesticides, just to cite a few. Some of the more current, and perhaps less familiar, additions include *xenobiotic transport, solute transport, Tiers 1 and 2, USEPA to cabinet status, and zero-discharge*. These are only the most prevalent topics of national interest. In more localized settings, residents are faced with leaking underground fuel tanks, movement of nitrates and industrial solvents into groundwater, air pollution and "stay-indoors" alerts in our major cities, radon seepage into homes, poor indoor air quality, chemical spills from overturned railroad tank cars, suspected health effects from living near high-voltage transmission lines, and food contamination by "flesh-eating" bacteria and other fungal or bacterial toxins.

It should then come as no surprise that the '90s generation is the first of mankind to have become afflicted with *chemophobia*, the pervasive and acute fear of chemicals.

There is abundant evidence, however, that virtually all organic chemicals are degraded or dissipated in our not-so-fragile environment, despite efforts by environmental ethicists and the media to persuade us otherwise. However, for most scientists involved in environmental contaminant reduction, there is indeed room for improvement in all spheres.

Environmentalism is the newest global political force, resulting in the emergence of multi-national consortia to control pollution and the evolution of the environmental ethic. Will the new politics of the 21st century be a consortium of technologists and environmentalists or a progressive confrontation? These matters are of genuine concern to governmental agencies and legislative bodies around the world, for many serious chemical incidents have resulted from accidents and improper use.

For those who make the decisions about how our planet is managed, there is an ongoing need for continual surveillance and intelligent controls to avoid endangering the environment, the public health, and wildlife. Ensuring safety-in-use of the many chemicals involved in our highly industrialized culture is a dynamic challenge, for the old, established materials are continually being displaced by newly developed molecules more acceptable to federal and state regulatory agencies, public health officials, and environmentalists.

Adequate safety-in-use evaluations of all chemicals persistent in our air, foodstuffs, and drinking water are not simple matters, and they incorporate the judgments of many individuals highly trained in a variety of complex biological, chemical, food technological, medical, pharmacological, and toxicological disciplines.

Reviews of Environmental Contamination and Toxicology continues to serve as an integrating factor both in focusing attention on those matters requiring further study and in collating for variously trained readers current knowledge in specific important areas involved with chemical contaminants in the total environment. Previous volumes of *Reviews* illustrate these objectives.

Because manuscripts are published in the order in which they are received in final form, it may seem that some important aspects of analytical chemistry, bioaccumulation, biochemistry, human and animal medicine, legislation, pharmacology, physiology, regulation, and toxicology have been neglected at times. However, these apparent omissions are recognized, and pertinent manuscripts are in preparation. The field is so very large and the interests in it are so varied that the Editor and the Editorial Board earnestly solicit authors and suggestions of underrepresented topics to make this international book series yet more useful and worthwhile.

Reviews of Environmental Contamination and Toxicology attempts to provide concise, critical reviews of timely advances, philosophy, and significant areas of accomplished or needed endeavor in the total field of xenobiotics in any segment of the environment, as well as toxicological implications. These reviews can be either general or specific, but properly they may lie in the domains of analytical chemistry and its methodology, biochemistry, human and animal medicine, legislation, pharmacology, physiology, regulation, and toxicology. Certain affairs in food technology concerned specifically with pesticide and other food-additive problems are also appropriate subjects.

Justification for the preparation of any review for this book series is that it deals with some aspect of the many real problems arising from the presence of any foreign chemical in our surroundings. Thus, manuscripts may encompass case studies from any country. Added plant or animal pest-control chemicals or their metabolites that may persist into food and animal feeds are within this scope. Food additives (substances deliberately added to foods for flavor, odor, appearance, and preservation, as well as those inadvertently added during manufacture, packing, distribution, and storage) are also considered suitable review material. Additionally, chemical contamination in any manner of air, water, soil, or plant or animal life is within these objectives and their purview.

Normally, manuscripts are contributed by invitation, but suggested topics are welcome. Preliminary communication with the Editor is recommended before volunteered review manuscripts are submitted.

Department of Entomology G.W.W.
University of Arizona
Tucson, Arizona

Table of Contents

Foreword v
Preface vii

Chlorpyrifos: Ecological Risk Assessment in North American Aquatic
Environments 1
    JOHN P. GIESY, KEITH R. SOLOMON, JOEL R. COATES, KENNETH R.
    DIXON, JEFFREY M. GIDDINGS, and EUGENE E. KENAGA

Cumulative and Comprehensive Subject Matter Index:
Volumes 151–160 131

Chlorpyrifos: Ecological Risk Assessment in North American Aquatic Environments

John P. Giesy, Keith R. Solomon, Joel R. Coats, Kenneth R. Dixon, Jeffrey M. Giddings, and Eugene E. Kenaga

Contents

I. Introduction	2
A. Problem Formulation	2
B. Endpoints	5
C. Analysis Plan	7
II. Stressor Characteristics	12
A. Physicochemical Properties	12
B. Mechanism of Action	13
III. Exposure Characterization	15
A. Use Patterns	15
B. Environmental Fate and Behavior	25
C. Measures of Exposure	27
IV. Effects Characterization	56
A. Bioconcentration and Bioaccumulation	56
B. Toxicity of Metabolites	58
C. Pulsed Exposures	59
D. Aquatic Ecotoxicology	59
E. Final Acute and Final Chronic Value Calculation for Chlorpyrifos	66

Communicated by George W. Ware
J. P. Giesy (✉)
Department of Zoology, National Food Safety and Toxicology Center, and Institute for Environmental Toxicology, Michigan State University, East Lansing, MI 48824, U.S.A.
e-mail: JGIESY@AOL.COM.

K. R. Solomon
Department of Environmental Biology, Centre for Toxicology, University of Guelph, Guelph, Ontario, N1G 2W1, Canada

J. R. Coats
Department of Entomology, Iowa State University, Ames, IA 50011, U.S.A.

K. R. Dixon
Institute of Environmental and Human Health, Texas Tech University, Lubbock, TX, 79409–1163, U.S.A.

J. M. Giddings
Springborn Laboratories, Wareham, MA 02571, U.S.A.

E. E. Kenaga
1584 E. Pine River Rd., Midland, MI 48640, U.S.A.

F. Toxicity of Sediment-Borne Chlorpyrifos 74
G. Microcosm and Mesocosm Studies with Chlorpyrifos 78
V. Risk Characterization 83
   A. Introduction 83
   B. Results of the Risk Analysis 93
VI. Ecological Significance of Effects 106
   A. Effects Criteria 106
   B. Ecological Role of Sensitive Taxa 107
   C. Spatial and Temporal Issues 109
   D. Return Frequency and Recovery of Affected Populations 109
   E. Population Modeling 110
VII. Uncertainties and Research Needs 113
Summary 115
Acknowledgments 120
References 120

I. Introduction
A. Problem Formulation

The objective of this risk assessment was to determine the probability and significance of effects of the organophosphate insecticide, chlorpyrifos, on aquatic ecosystems in North America. The assessment addressed both agricultural and nonagricultural uses. However, the primary focus of the risk assessment was agricultural ecosystems, especially row crops and, in particular, the "corn-belt" agroecosystems. The risk assessment also addressed potential effects from other agricultural uses as well as urban uses such as turf, termiticide, and home use. Exposure and effects in freshwater and saltwater environments were considered. Aquatic invertebrates and fish were included in the assessment, but amphibians, reptiles, birds, and mammals were not. The potential exposure of these organisms is small because of a lack of biomagnification of chlorpyrifos. Thus, if their prey are not affected, it is unlikely that organisms at higher trophic levels would be adversely affected. Measurements of chlorpyrifos residues in fish have shown both low probability and low concentrations of exposure (USEPA 1992b). Insufficient data on amphibians were available for a direct assessment of risks. A risk assessment of chlorpyrifos in terrestrial ecosystems was conducted in parallel with this aquatic risk assessment (Kendall et al., in manuscript). Chlorpyrifos is not used in isolation, and residues of other substances with the same mechanism of action may co-occur with chlorpyrifos and some of these may display additive toxicity (Bailey et al. 1997). Although the presence of these compounds could influence the overall conclusions of a risk assessment for the class of anticholinesterase insecticides, the extensive resources necessary to conduct a classwide review were not available and they were excluded from this evaluation.

Conceptual Model. Chlorpyrifos is an organophosphate insecticide used on agricultural crops, on turf, and to control termites and household pests. Each type of use may result in release to the environment. Agricultural use may lead to non-point source pollution through runoff and drift. Turf, termiticide, and other noncrop use may lead to contamination of urban and rural storm water runoff (point source or non-point source pollution). Household use (especially improper disposal) may lead to contamination of municipal sewage (point source discharge). Each of these pathways can transport chlorpyrifos to streams, rivers, lakes, ponds, wetlands, estuaries, or other aquatic ecosystems (Fig. 1). The major focus of this assessment is on the potential effects of chlorpyrifos in freshwater systems. It is recognized that estuarine ecosystems are the interface between freshwater and marine systems and that they also may become contaminated with chlorpyrifos. Dilution in estuaries would likely result in concentrations less than those in rivers and streams and the risk assessments in the latter would be protective of estuaries. Actual measurements from such systems were sparse in the literature; however, McConnell et al. (1997) recently reported measuring chlorpyrifos concentrations in Chesapeake Bay. The spatial and temporal distribution of chlorpyrifos use in agricultural crops, on turf, and to control termites and household pests was the key to verifying that concentrations in surface waters were relevant to the risk assessment.

The potential effects of chlorpyrifos exposure are shown in Fig. 2. Aquatic organisms vary in their sensitivity to chlorpyrifos. Aquatic plants are unaffected by chlorpyrifos at concentrations many times greater than those that occur in aquatic systems. Some species of fish and invertebrates are extremely sensitive, while others are relatively tolerant. Chlorpyrifos could therefore lead to a reduction or elimination of some invertebrate populations, and possibly cause fish kills or chronic effects on fish growth or reproduction. It is possible that some fish may also be affected indirectly through reduction in their food supply. The

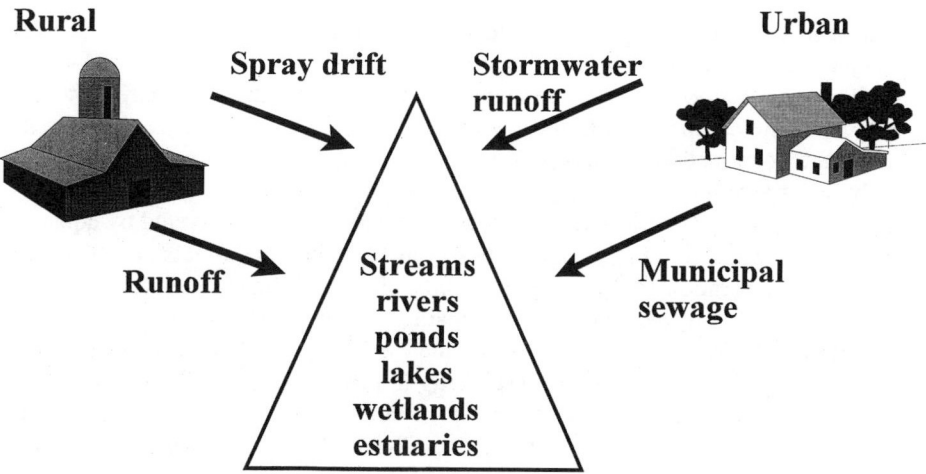

Fig. 1. Conceptual model: possible transport pathways of chlorpyrifos to aquatic ecosystems.

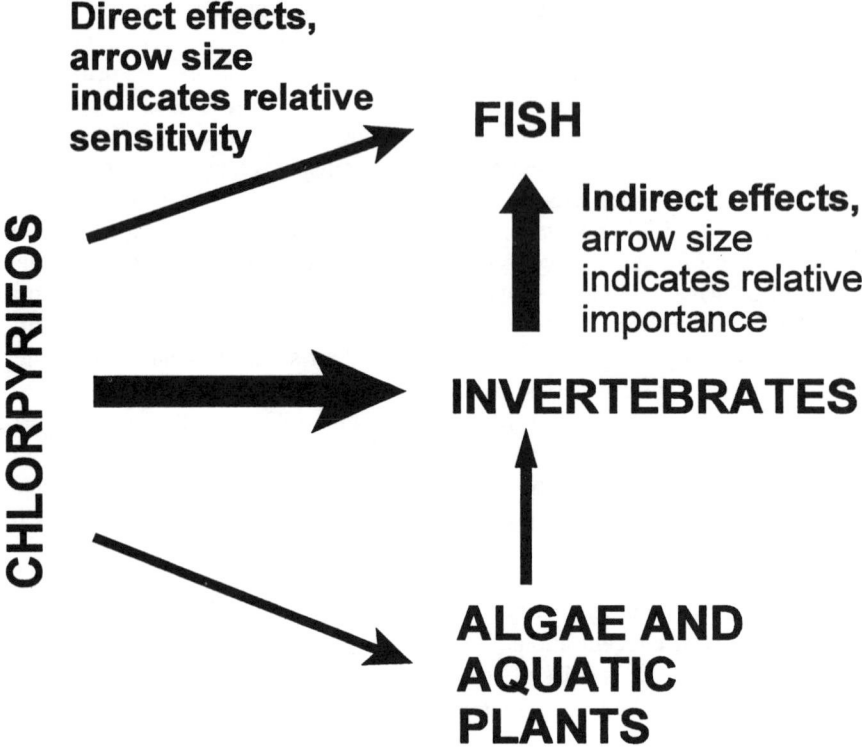

Fig. 2. Conceptual model: potential effects of chlorpyrifos on ecological receptors.

extent of such direct and indirect effects, and their ecological significance, were subjects of investigation in this risk assessment.

The approach to the risk assessment is determined by the characteristic toxic action of the compound and the nature of the aquatic exposure. As discussed here, the acute action of chlorpyrifos and other organophosphorus insecticides is by inactivation of acetylcholinesterase enzyme activity (AChE). Although the toxic action is rapid, chlorpyrifos is readily metabolized and excreted by aquatic organisms, and after chlorpyrifos exposure is removed, the effects are reversed through regeneration of AChE. The toxicity of chlorpyrifos is therefore not cumulative under the frequencies and intensities of most environmentally relevant exposures (Giddings et al., 1997; Jarvinen et al. 1988; Macek et al., 1972; Naddy 1996). Exposure patterns, especially in lower-order streams where chlorpyrifos concentrations tend to be greatest, consist of sharp pulses associated with individual runoff or spray drift events. Typically, the duration of individual exposure pulses in streams is of the order of 2 d. Therefore, this risk assessment focused on comparison of acute toxicity (48-hr LC_{50}s) with 48-hr time-weighted average exposure concentrations. To be conservative, in cases in which sampling frequency did not allow estimation of 48-hr time-weighted averages, 48-hr LC_{50}s were compared with instantaneous concentration measurements that would greatly overestimate the potential for adverse effects.

The potential ecological effects of chlorpyrifos may be modified by various factors, including spatial/temporal distributions of exposure and of receptors, ecological interactions between populations, ecological recovery mechanisms, water quality and other physical and chemical conditions, and natural variability. Some of these factors may mitigate the impact of chlorpyrifos, while others may augment the severity of effects. Because most of these factors tend to mitigate exposures, to be conservative, the assessment presented here did not address these issues.

Negligence, accidents, and misuse of chlorpyrifos are potential sources of chlorpyrifos entry into surface waters. Because these are episodic events that are difficult to subject to detailed analysis or modeling, the exposures and effects resulting from such occurrences were judged to be outside the scope of this risk assessment. Because of a lack of exposure data specifically related to landscape and structural uses of chlorpyrifos, it was not possible to conduct formal risk assessments for these uses. However, some studies have sampled aquatic systems in industrial and urban areas, and nonspecific concentration data for uses in these environments were available for analysis.

B. Endpoints

Assessment Endpoints. Assessment endpoints are "explicit expressions of the actual environmental value that is to be protected" (USEPA 1992a). The endpoints for this risk assessment were selected on the basis of the pathways of chlorpyrifos exposure, the patterns of chlorpyrifos toxicity, and judgments about the ecological, economic, and social importance of ecosystem components at risk.

Each assessment endpoint incorporates two elements: a valued ecological entity and a characteristic of that entity (USEPA 1996a). The assessment endpoints for the chlorpyrifos aquatic risk assessment are shown in Table 1. Because the risk assessment addresses a broad range of ecosystem types and exposure patterns, the assessment endpoints were necessarily generic. The focus of these assessment endpoints was on populations and communities, not individual

Table 1. Assessment endpoints for chlorpyrifos aquatic risk assessment.

Entity	Characteristic(s) to be protected
Fish	Population persistence (a function of survival, growth, and recruitment)
Invertebrates (benthic and planktonic)	Community productivity

organisms, but this does not mean that the survival of individuals is not important. Rather, because a probabilistic approach was used, the experimental unit on which effects were assessed was the individual organism. These responses were expressed on a population rate basis. For instance, individual fish can die, but the response reported is mortality, which is an aggregate response of population-level mortality rates. Individuals cannot have "rates of mortality." In this assessment, death of an individual fish was not considered an adverse or undesirable effect unless it belonged to an endangered species. However, reduced survival, growth, or reproduction of many fish in a population would be of concern if it resulted in a long-term change in population size, structure, or distribution. Widespread and repeated mortality (fish kills) would also be an undesirable effect.

For most invertebrates, even changes at the population level may not be ecologically, economically, or socially significant, except in the case of endangered species, if the overall productivity of the invertebrate community and its ability to support higher trophic levels are maintained. Reduction or loss of populations of numerically minor or ecologically redundant invertebrate species may have little or no impact on the rest of the ecosystem.

The assessment endpoints selected do not explicitly address ecosystem structure or function. Ecosystem integrity will be protected if fish populations and invertebrate communities are protected. Because of their extreme lack of sensitivity, plants would be unlikely to be directly affected by chlorpyrifos (Barron and Woodburn 1995). Because of their structural and functional redundancy, ecosystems are less sensitive to chemical impacts than are their component populations and communities. In this risk assessment, population-level consequences of reductions in individual survival, growth, and reproductive success were assessed. The significance of changes in invertebrate populations in terms of overall community production and dietary requirements of fish were also addressed.

Measures of Effect. Measures of effect are the parameters that are actually quantified as indicators of the effect of the stressor (USEPA 1996a) and are ultimately linked to assessment endpoints. EPA's previous Guidelines for Ecological Risk Assessment (USEPA 1992a) refer to these as "measurement endpoints." In risk assessments of chemicals, the most commonly used measurement endpoints are survival and sometimes growth or reproduction of individual organisms, as determined in laboratory toxicity tests. Acute toxicity, which is usually expressed as a median lethal value (LC_{50}, the concentration of toxicant in water or sediment that causes death to 50% of the test organisms exposed), is particularly relevant in the case of chlorpyrifos because its toxic effects are rapid and exposures are of relatively short duration in surface waters. Chronic toxicity, usually expressed as a no-observed-effect concentration (NOEC, the greatest concentration of toxicant that causes no statistically significant change in survival, growth, or reproduction of the test organisms exposed), while important, is of secondary concern. This is because chlorpyrifos has a relatively small acute-to-chronic ratio and does not persist in surface waters (Barron and Woodburn 1995; Racke 1993). For this risk assessment, acute toxicity values for freshwater or saltwater organisms (fish and invertebrates) constituted the most important set of measurement endpoints. To be conservative, chronic toxicity, either measured directly or extrapolated from acute toxicity values, was also considered.

In addition to simple, single-species toxicity tests conducted under laboratory conditions, chlorpyrifos has been extensively studied in situations that permit direct measurements of effects on populations and communities. These include field studies and mesocosm or microcosm experiments. Mesocosms and microcosms are small physical models of ecosystems, usually including both sediment and water, and always incorporating interactions among organisms representing at least two trophic levels. Because results of such studies are available for chlorpyrifos, the measurement endpoints for this risk assessment included invertebrate population densities, invertebrate community composition, and fish population structure. These measurement endpoints are more relevant and predictive than laboratory toxicity data with respect to the assessment endpoints. However, they are also more difficult to evaluate and interpret because of the complexity and variability of ecosystem-level experimental systems, and professional judgment is needed to integrate them into the risk assessment.

The risk assessment used a lines-of-evidence approach and expert judgment to integrate laboratory toxicity measurements and microcosm results into a reference value for ecological effects of chlorpyrifos. The laboratory toxicity data were used to support a quantitative probabilistic assessment, and the microcosm studies supported a more qualitative evaluation of ecological significance.

C. Analysis Plan

The following sections outline our approach to evaluating the data and reaching conclusions about the probability and significance of ecological effects of chlorpyrifos in aquatic systems. The analysis consisted of four parts: (1) stressor characterization, a review of the inherent characteristics of chlorpyrifos; (2) analysis of potential exposure by various pathways; (3) analysis of effects on single species, populations, and communities; and (4) risk characterization (Fig. 3).

Characteristics of the Stressor. Information about the inherent properties and characteristics of chlorpyrifos was reviewed. This information included physical properties, chemical properties, mechanism of toxic action, specificity of the toxic mechanism, reversibility of toxic effect, and reciprocity (the relationship between concentration, exposure duration, and toxic effect). The review also addressed the characteristics of known chlorpyrifos transformation products, especially 3,5,6-trichloro-2-pyridinol (TCP). Basic properties of chlorpyrifos formulations were summarized and used to select and scale measurement endpoints.

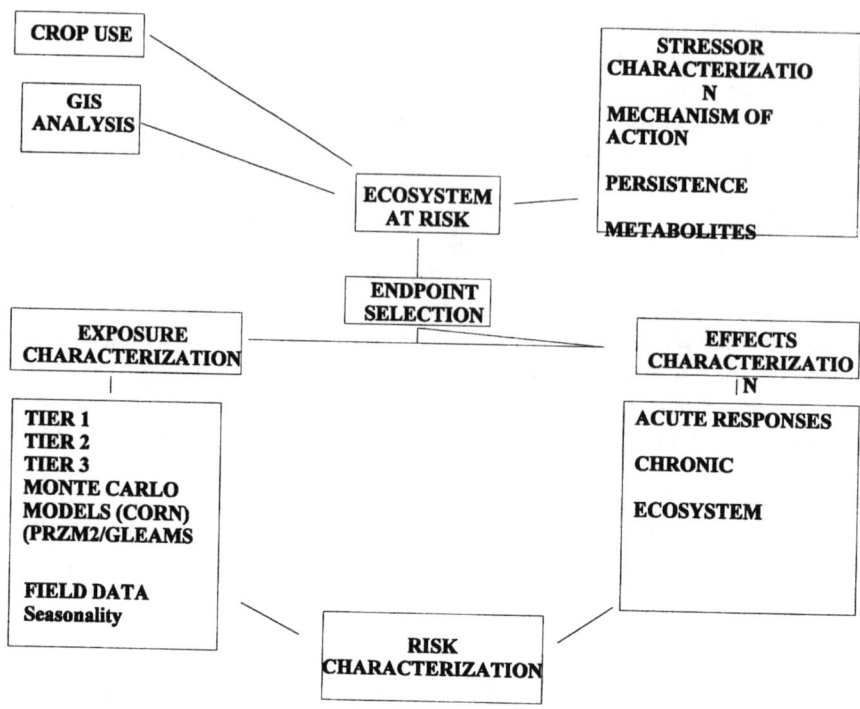

Fig. 3. The process followed in the chlorpyrifos risk assessment for aquatic ecosystems.

Exposure Analysis. The objective of the exposure analysis was to determine the concentrations of chlorpyrifos in water and sediment of aquatic ecosystems over time. The analysis began with a review of the patterns of chlorpyrifos use, including agricultural and nonagricultural applications. Geographic Information System (GIS) techniques were used to relate the quantities and timing of chlorpyrifos use to the distribution of aquatic habitats and ecological regions, and thereby to focus further investigation on ecosystems where exposure is likely to be the greatest.

The exposure analysis then reviewed information on the fate of chlorpyrifos in terrestrial and aquatic environments. Much of this information had already been compiled by Racke (1993). Environmental compartments and pathways considered included persistence in soil, water, and sediment; biotic and abiotic degradation pathways and degradation products; partitioning between solids, water, organic matter, and air; and factors influencing bioavailability in water and sediment. This information helped refine the understanding of exposure pathways and the influence of environmental factors on exposure. Some of the quantitative information was also used as input to exposure models.

Risk assessments can be conducted in both *a priori* and *a posteriori* contexts. If no information on the distribution of the compound of interest in the environment is available, the exposure component of the risk assessment must be estimated. This is done by the use of simple "worst-case" scenarios in Tier I assessments. The risk assessment can be refined by the use of mathematical models to simulate concentrations of chemicals in various environmental compartments. Alternatively, in the case in which a chemical has been in use, instead of estimating environmental concentrations, actual measured values from monitoring programs can be used to estimate risk. This is the case for chlorpyrifos, where use was made of the relatively extensive monitoring information on chlorpyrifos to determine the probability of exceeding a specified hazard value. The authors recognize that environmental monitoring data could be subject to biases resulting from the fact that these data sets are relatively short (<10 yr) and that sampling was not necessarily designed to specifically address distributional analysis. Nevertheless, the data for chlorpyrifos were judged to be extensive and complete enough to allow a realistic exposure profile to developed.

Estimates of exposure concentrations were developed using two parallel approaches: simulation modeling, and analysis of surface water monitoring data. The modeling addressed runoff from the most important agricultural use of chlorpyrifos. The modeling effort proceeded through three phases, or tiers, consistent with the recommendations of the Aquatic Risk Assessment and Mitigation Dialog Group (SETAC 1994). Tier I was a simplistic, generic calculation of concentrations in water running off from agricultural fields after a single application of chlorpyrifos, using EPA's GENEEC model (Parker and Rieder 1995). Tier II involved simulation of two representative sites, selected to represent extreme and median environmental conditions conducive to runoff of chlorpyrifos over multiple years, using the PRZM (Mullins et al 1993) runoff model and the EXAMS water quality model (Burns 1990). In Tier III, multiple sets of input parameters for the GLEAMS runoff model and EXAMS were selected relative to their respective probability distributions, and the model output was, in turn, presented as probability distributions of chlorpyrifos concentrations.

The second approach to characterizing patterns of chlorpyrifos exposure was to analyze available data on concentrations of chlorpyrifos in several North American watersheds. Several comprehensive data sets that include results of chlorpyrifos analyses have been compiled by various agencies and individual researchers. These were examined for quality, and those considered reliable became the basis for statistical summarization and exploration of trends. As with many data sets for pesticides in surface waters, the focus for sampling was directed to the time of the year when major use occurs. Using these data sets as representative of the entire year may introduce a conservative bias toward higher frequencies of occurrence; however, in the corn production areas, the use season coincides with the season of greater biological activity in freshwater systems. Therefore, the coincidence of seasonal exposure with seasonal biological activity actually adds additional realism to the assessment. As has been observed for other pesticides (SETAC 1994; Solomon et al. 1996; Solomon and Chappel 1997), measured concentrations of chlorpyrifos were expected to approximate log-normal or similar nonsymmetrical distributions. Point estimates from the distributions of chlorpyrifos concentrations were compared with benchmarks of toxicity and effects in the risk characterization phase of the assessment.

Empirical measures of exposure can be expressed as time-weighted mean concentrations (TWMCs) or as instantaneous measured concentrations. For some of the larger data sets, it was possible to calculate TWMCs for a time interval of 48 hr, moving the window 1 day at a time, adding the daily concentrations, and then calculating the time-weighted mean concentration by dividing the total by two. In cases where the data were not available, the instantaneous maximum was assumed to have a duration of 48 hr. This assumption made the assessment more conservative (protective). The exact degree of conservatism is a function of the duration of exposure pulses and difference between the duration of the maximum concentration pulse width and 48 hr. In all cases, however, the estimates of risk presented are overestimates as the result of the assumption of a square-wave pulse during which the maximum concentration was maintained for the entire pulse.

Effects Analysis. The toxicity of chlorpyrifos has been the subject of a number of laboratory studies (Barron and Woodburn 1995), and results of these studies were the basis for quantitative analysis of potential effects. A wide range of data was available for arthropods and fish but data for amphibians were few. Those data that were available suggested that amphibians were of similar sensitivity to fish. Thus, while risks could not be specifically addressed, the assessment of fish was judged to be applicable to the juvenile stages of amphibians. Like the surface water concentration measurements, the toxicity data (LC_{50}/EC_{50} values) were assumed to be described by log-normal distributions. Data were examined for reliability, and then used to calculate distribution parameters for various subsets of test species: fish, arthropods, freshwater, saltwater, and all various combinations of these. Various probabilities of response (proportion of species responding) were examined. In general, the 10th centile of each distribution (i.e., the concentration at which 10% or less of the test species would be affected) was compared with exposure distributions in the risk characterization phase of the assessment. Chronic toxicity data were reviewed to derive an average acute-to-chronic ratio (ACR), which was then used to derive a reference value for chronic toxicity from the reference values for acute toxicity.

An attempt was made to explicitly relate the toxicity results (measurement endpoints) to effects on populations (assessment endpoints) using ecological models. Data on survival and growth were used in population life table models (e.g., Foran et al. 1991; Meyer et al. 1987) to assess potential effects of chlorpyrifos on population dynamics.

The effects analysis also included a review of results of field experiments and mesocosm and microcosm studies of chlorpyrifos. Ecosystem-level experiments allow observation of the responses of populations and communities under quasi-natural conditions, including competitive and predator–prey interactions. Observations in experimental ecosystems are useful for calibrating the results of laboratory studies (Giddings and Franco 1985) and for interpreting the ecological significance of effects predicted by the risk characterization. Effects on fish populations and invertebrate communities in experiments with chlorpyrifos were measurement endpoints in the risk assessment that were directly linked to the assessment endpoints (see Table 1).

Risk Characterization. Risk characterization integrated the results of the exposure analysis and the effects analysis to estimate the risk of impacts on the assessment endpoints (see Table 1). This risk assessment of chlorpyrifos followed the guidance of the Aquatic Risk Assessment and Mitigation Dialog Group (SETAC 1994) by expressing risk as the probability of exposure exceeding a specific ecological effect measure, such as the 10th centile of acute toxicity values, which was used as a benchmark. This probability was estimated from the concentration distribution derived from the exposure analysis. A probability of effect can be estimated for each set of exposure data, such as for specific sites or groups of sites, and for specific time periods, and for each ecological effect criterion, such as acute or chronic toxicity to specific groups of organisms or a no-observed-adverse-effect concentration (NOAEC) based on mesocosm and microcosm studies. No explicit estimates of the degree of overlap between probability of exposure and response were calculated. For the purposes of this risk assessment, the 10th centile of the toxicity distribution was used as a reference point. Although this synthetic parameter has no direct ecological interpretation, it was calibrated against ecosystem-level and chronic responses and it does put into context the general degree of effect that would be expected at a particular exceedence of a response concentration and was useful for assessing priority based on risks rather than hazards. The risk was further characterized by determining which species would be predicted to be affected, the likelihood that the potentially affected species would occur in the ecosystem of interest, and the degree of "functional importance" of the species in that system.

In situations where the probabilistic risk analysis indicated substantial risk, further consideration was given to factors that influence risk in specific situations. One aspect examined was the co-occurrence in space and time of great exposures and sensitive receptor organisms. Exposures with the greatest probability of exceeding effect criteria were identified, and these specific situations were studied to determine where, when, and for how long potentially hazardous exposures occur and which populations are present at those places and times. The relationship between exposure concentration, exposure, duration, and toxic effect (reciprocity) was also examined to refine the assessment of risk in these situations.

Another factor considered was the ecological role of the most sensitive organisms in the specific ecosystems at greatest risk. This required an understanding of food webs, life histories, demographics, and population dynamics in the ecosystems of concern. This information was used in a qualitative manner to support informed professional judgements about the seriousness of potential ecological effects.

II. Stressor Characteristics

Chlorpyrifos [O,O-diethyl O-(3,5,6-trichloro-2-pyridyl) phosphorothioate)] is a widely used, broad-spectrum organophosphorothioate pesticide that displays activity against a broad range of insect pests and is used in a wide variety of global markets. Agricultural, industrial, and residential uses of chlorpyrifos have resulted in intentional and accidental introduction of the compound into an array of terrestrial and aquatic ecosystems throughout the world. An extensive database on the toxicology of chlorpyrifos to numerous aquatic and terrestrial organisms has developed since its discovery (in 1962) and the initial production of the chemical in 1965. A brief summary of the physicochemical properties and mechanisms of action of chlorpyrifos is presented next.

A. Physicochemical Properties

The available data on the physical and chemical properties of chlorpyrifos have been reviewed in detail by Racke (1993) and are summarized in Table 2. Chlorpyrifos is an organophosphorothioate compound. Its water solubility is 1.4 mg/L at 25 °C (Packard 1987), and it exhibits a moderate level of hydrophobicity (log K_{OW} of 4.7–5.3; De Bruijn et al. 1989; McDonald et al. 1985). As a result of its hydrophobicity, chlorpyrifos partitions extensively from the aqueous phase

Table 2. Summary of physical and chemical properties of chlorpyrifos.

Property	Value[a]
CAS number	2921-88-2
Chemical name	O,O-Diethyl$O$-(3,5,6-trichloro-2-pyridyl)-phosphorothioate
Molecular weight	350.6
Molecular formula	$C_9H_{11}NO_3PSCl_3$
Melting point	41°–44°C
Water solubility (distilled)	1.39 mg/L at 25°C
Vapor pressure	2.0×10^{-5} mm Hg at 25°C
Henry's law constant	6.64×10^{-3} atm-L mol^{-1}
Log K_{OW}	4.7–5.3
Hydrolysis (25°C)	Below pH 7, average $t_{1/2}$ of 77 d Above pH 7, average $t_{1/2}$ 10–16 d
Aqueous photolysis	Average $t_{1/2}$ 30 d under midsummer sunlight at ~40°N latitude
Soil photolysis	Wide range of values, from no observed degradation to $t_{1/2}$ 17 d on moist soil
Aerobic soil metabolism (25°C)	Variable, average $t_{1/2}$ 30–60
Anaerobic aquatic metabolism (25°C)	Variable $t_{1/2}$, 40–50 d to 150 d

CAS, Chemical Abstracts Service.
[a]From Racke (1993).

into the organic fractions of environmental matrices. Chlorpyrifos is strongly adsorbed by soil and sediment, and it displays a mean sorption coefficient (K_{OC}) of about 8500 mL/g (Racke 1993). Sorptive equilibrium in soil–water systems is reached quickly, generally within 2–4 hr (Felsot and Dahm 1979; McCall 1987). Although chlorpyrifos has an intermediate vapor pressure (2×10^{-5} mm Hg at 25 °C; Chakrabarti and Gennrich 1987), volatilization has been shown to be a significant mechanism of dissipation from certain environmental surfaces (i.e., plant foliage, pond water), as summarized by Racke (1993).

B. Mechanism of Action

The toxicity of chlorpyrifos results from initial metabolic activation to form chlorpyrifos oxon, with subsequent inactivation of acetylcholinesterase (AChE) at neural junctions; the inactivation of AChE often occurs by oxon phosphorylation of the enzyme active site (Fig. 4). In some fish species, AChE can become irreversibly inhibited through dealkylation of the phosphorylated AChE, a process that renders it resistant to hydrolysis (Chambers and Chambers 1989). AChE inactivation is dose- and duration dependent, and may result in overstimulation of the organism's peripheral nervous system and subsequent toxicity.

Upon exposure to sublethal concentrations of organophosphorus (OP) insecticides, inhibition of brain and muscle AChE has been observed within 24 hr, which indicates that bioactivation occurred rapidly after exposure (Benke et al. 1974). The failure of brain AChE to spontaneously regenerate in some fish species indicates that dephosphorylation of the AChE enzyme does not generally occur and that restoration of AChE activity requires synthesis of new AChE or

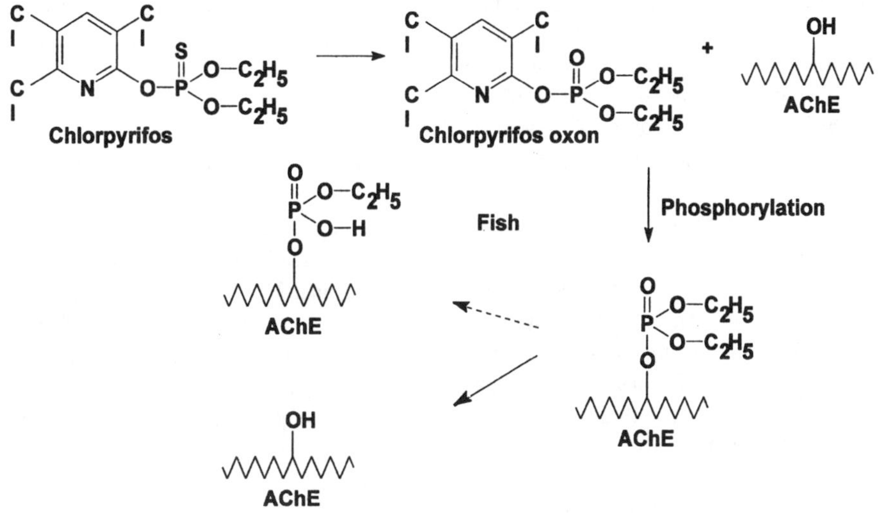

Fig. 4. Diagram of activation of chlorpyrifos to the oxon form, phosphorylation, and recovery of acetylcholinesterase (AChE).

the usage of sequestered AChE not previously active or exposed to the OP oxon (Boone and Chambers 1996; Weiss 1961). The time required to regenerate brain AChE to normal activity depends on the extent of initial inhibition, the OP compound, and the particular species; a lack of spontaneous reactivation has been observed with rainbow trout (*Oncorhyncus mykiss* Walbaum), while fathead minnows (*Pimephales promelas*) have demonstrated recovery of AChE to control activities within 2 wk of termination of OP exposure (Wallace and Herzberg 1988; Weiss 1961). In general, AChE inhibition of >50% in fish brain tissue may require more than 4 wk for AChE activities to recover to concentrations seen in control samples (Boone and Chambers 1996; Weiss, 1961).

In this assessment, the reciprocity between duration and intensity of exposure was investigated. The intent was to support the selection of appropriate durations for comparing measures of exposure and response and to determine the effect of response frequency on interpreting exposure, that is, what is the potential effect of multiple pulsed exposures, what is the minimum interpulse period required before pulses can be considered to be independent events, and what is the minimum interpulse duration for exposures to be simply additive? To be conservative, when comparing 2-d TWMCs, it was assumed that there would be no regeneration of AChE activity.

It has been proposed that some fish have a greater reserve of AChE than standard mammalian test species such as rats and should therefore be able to withstand a greater degree of inhibition, assuming that a minimum amount of AChE is needed to sustain life (Boone and Chambers 1996; Johnson and Wallace 1987). Indeed, at the LD_{50}/LC_{50} dose for malaoxon and paraoxon, 50%–70% of the rat brain AChE was inhibited, whereas in the rainbow trout, 70%–90% of brain AChE was inhibited; i.e., a greater degree of enzyme inhibition is associated with mortality in fish compared to rats (Johnson and Wallace 1987). It has been observed that >85% inhibition of sheepshead minnow (*Cyprinodon variegatus*) brain AChE activity was required to cause 24- to 96-hr acute toxicity with various OP insecticides (Coppage 1972). The greater amount of AChE inhibition that can occur in surviving fish suggests that they do not require as much functional AChE to sustain life as do mammals (Boone and Chambers 1996).

Species differences in behavior, feeding ecology, receptor sensitivity, and pharmacokinetics result in a greater than 1 million fold variation in sensitivity to chlorpyrifos among species (Barron and Woodburn 1995; Marshall and Roberts 1978). Specifically, individual and species susceptibility to chlorpyrifos is related to the binding affinity of chlorpyrifos oxon to AChE and to its subsequent rate of inactivation. Affinity of other esterases such as aliesterases (AliE) to associate with OP oxons may also provide some protection against AChE inhibition and may also help explain the interspecies variability of acute toxic responses to chlorpyrifos and other OPs (Boone and Chambers 1996). More rapid inhibition of mosquitofish (*Gambusia affinis*) AliE than that of brain or muscle AChE following chlorpyrifos exposure has been observed, which suggests that the greater AliE affinity for chlorpyrifos oxon helped afford the animal substantial protection from toxicity (Boone and Chambers 1996). However, this capacity may not be present in all fish nor equally effective in all life stages. In contrast, the limited inhibition of AliE in mosquitofish following methyl parathion exposure indicated that mosquitofish AliE has lesser affinities for methyl parathion oxon and would afford less protection of AChE. In addition, differences in the strength of the oxon–AChE association can play a significant role in the widely variable OP potencies exhibited by different species (Carr and Chambers 1996). Finally, it has been suggested that variations in conjugation of OP compounds with glutathione may also help explain species-specific OP susceptibility; glutathione conjugation with xenobiotics is generally considered protective in that it facilitates xenobiotic excretion (Hasspieler et al. 1994; Motoyama and Dauterman 1980).

III. Exposure Characterization
A. Use Patterns

Agricultural Uses. The major uses of chlorpyrifos are in both agricultural and urban pest control markets. Since initial commercialization of chlorpyrifos in the mid-1970s in corn, cotton, and peaches, agricultural use of chlorpyrifos has expanded to include additional crops (Tables 3 and 4). A variety of foliar, dormant tree, and soil applications are utilized to control numerous insect pests. Use in field corn represents the largest agricultural market. The most common treatment in field corn is a T-band application of the 15% a.i. granular (15G) formulation at planting time. Granular applications also are important in the next largest application, which is to peanuts. Most of the other crop uses employ dilute sprays of the 4 lb a.i./gal emulsifiable concentrate (4E) formulation. The (50% Wettable Powder) (50W) formulation is used primarily in pome fruits and, to a lesser extent, in nuts. Application rates vary by use, but for most scenarios a single treatment is typically applied annually.

Urban Uses. Chlorpyrifos is applied extensively in nonagricultural situations (Table 4). The major urban market is structural wood protection, where the greatest volume of use is soil treatment to create a barrier against termite infestation. Trenching or rodding applications along the perimeter of the structure utilize relatively concentrated dilutions of emulsifiable formulation (typically 0.75%–1% of a 4E formulation) injected in sufficient volume to form a local barrier. Wood protection treatments are applied by trained and licensed professional applicators. Retail products purchased and applied by residents or workers in homes and businesses make up the next largest volume of use. These uses involve both indoor and outdoor applications of a large number of different formulations. The other uses are professional uses in and around dwellings and commercial businesses, again employing a great variety of formulated products (Table 4).

Table 3. Agricultural uses of chlorpyrifos.

Crop	Percent of total use in crop by formulation			Percent of total use in all crops	Key pests	Typical use rate[b] (lb a.i.)	Application methods
	4	15 G	50 W				
Field corn	8	92	0	55	Corn rootworm larvae	1.0–1.2	15G[c]: Soil (T-band); 4E[d]; Soil (PPI[e]), chemigation
					Cutworms	1.0–1.2	15G: Soil (T-band); 4E: Soil (PPI), Foliar (ground, air)
					Wireworms/grubs	1.3–2.0	15G: Soil (T-band, In-Furrow); 4E: Soil (PPI)
					European corn borer	1.0	15G 4E: Foliar (ground, air); 4E: Foliar (chemigation)
					Corn rootworm beetles	0.5 (1–2)	4E: Foliar (air, chemigation)
Peanut	1	99	0	7	Lesser corn stalkborer	2.0	15G: Soil (band), Foliar (band)
					Southern corn rootworm	2.0	15G: Soil (band), Foliar (band)
					Cutworms	2.0	15G: Soil (band)
					White mold	2.0	15G: Soil (band), Foliar (band)
Citrus	97	3	0	6	California red scale	6.0	4E: Foliar (airblast)
					Lepidopteran larvae	2.0	4E: Foliar (airblast)
					Citrus rust mite	3.5	4E: Foliar (airblast)
					Fire ants	1.0 (3)	15G: Soil (surface); 4E: Soil (surface, on fertilizer)
					Mealybugs	2.0	4E: Foliar (airblast)
Alfalfa	100	0	0	6	Alfalfa weevil	0.75	4E: Foliar (ground, air)
					Aphids	0.75	4E: Foliar (ground, air)
					Armyworms	1.0	4E: Foliar (ground, air)
					Potato leafhopper	0.5	4E: Foliar (ground, air)

Crop				Pest	Rate	Formulation: Application
Cotton	100	0	0	5	Aphids	0.75	4E: Foliar (ground, air)
					Plant bugs	0.33	4E: Foliar (ground, air)
					Armyworms	1.0	4E: Foliar (air)
					Pink bollworm	1.0 (1.5)	4E: Foliar (ground, air)
					Salt marsh caterpillar	1.0	4E: Foliar (ground, air)
Nuts (foliar)	86	0	14	5	Codling moth	2.0	4E, 50W[f]: Foliar (airblast)
					Walnut husk fly	2.0	4E, 50W: Foliar (airblast)
					Walnut scale	2.0	4E, 50W: Foliar (airblast)
					Navel orangeworm	2.0	4E, 50W: Foliar (airblast)
					Peach twig borer	2.0	4E, 50W: Foliar (airblast)
					San Jose scale	2.0	4E, 50W: Foliar (airblast)
					Ants	2.0	4E, 50W: Soil (ground, orchard floor)
					Pecan nut casebearer	2.0	4E, 50W: Foliar (airblast)
					Hickory shuckworm	2.0	4E, 50W: Foliar (airblast)
Tobacco	100	0	0	3	Cutworms	3.0	4E: Soil (PPI)
					Wireworms	3.0	4E: Soil (PPI)
					Mole crickets	3.0	4E: Soil (PPI)
Sugarbeet	73	27	0	3	Sugarbeet root maggot	0.5	4E: Foliar (ground, air)
					Cutworms	1.0	4E: Foliar (ground, air)
					Beet armyworm	1.0	4E: Foliar (ground, air)
					Grasshoppers	0.25	4E: Foliar (ground, air)
					Sugarbeet root maggot	1.8	15G: Soil (band)
Dormant trees	100	0	0	3	San Jose scale	2.0	4E: Airblast
					Oblique banded leafroller	2.0	4E: Airblast
					Rosy apple aphid	2.0	4E: Airblast
					Pandemis leafroller	2.0	4E: Airblast
					Peach twig borer	2.0	4E: Airblast

(Continued)

Table 3. Continued.

Crop	Percent of total use in crop by formulation			Percent of total use in all crops	Key pests	Typical use rate[b] (lb a.i.)	Application methods
	4	15 G	50 W				
					Brown almond mite	2.0	4E: Airblast
					Pear psylla	2.0	4E: Airblast
					Climbing cutworms		
Pome fruit (foliar)	0	0	100	2	Rosy apple aphid	1.25	50W: Foliar (airblast)
					Oblique banded leafroller	1.25	50W: Foliar (airblast)
					Tufted apple bud moth	1.25	50W: Foliar (airblast)
					Pandemis leafroller	1.5	50W: Foliar (airblast)
					San Jose scale	1.5	50W: Foliar (airblast)
Sorghum	100	0	0	1	Greenbug	0.5	4E: Foliar (ground, air)
					Sorghum midge	0.25	4E: Foliar (ground, air)
wheat	100	0	0	1	Russian wheat aphid	0.38	4E-SG[g]: Foliar (air)
					Greenbug	0.38	4E-SG: Foliar (air)
					Grasshoppers	0.25	4E-SG: Foliar (air)
					Western bean cutworm	0.5	4E-SG: Foliar (air)

[a] 1991–1993 average.
[b] Typical number of applications is one, except where noted within parentheses. Use rates are given in pounds per acre to maintain consistency with chlorpyrifos labels.
[c] 15% a.i. granular formulation.
[d] 4 lbs a.i. per gallon emulsifiable concentrate formulation.
[e] Preplant incorporated.
[f] 50% a.i. wettable powder formulation.
[g] 50% a.i. soluble granular formulation.
Use rates are given in pounds per acre to maintain consistency with chlorpyrifos labels.

Table 4. Urban uses of chlorpyrifos.

Uses	Percent of total use	Major uses	Typical use rate	Typical spray/application volume[c]	Typical application frequency (per year) Professional	Typical application frequency (per year) Retail
Wood protection (home)	70	Subterranean termites	0.75–1%	192 gal./structure (soil)	Once in 7–9 yr	—
	14	Cockroaches/ants	0.25–0.5%	0.25–0.5 gal./structure	—	1–2
		Fleas	0.25–0.5%	0.5–0.75 gal./structure	—	1–2
		Pet collars	Cats 3%, Dogs 8%	—	—	1–2
		Building foundations (2–3 ft)	0.2–0.5%	1–2 gal./structure	1	—
		Around structures (6–10 ft)	0.03–0.06%	10–15 gal./structure	1	—
		Turf (liquid)	1 lb. a.i./A	2–4 gal./1000 sq. ft.	1	—
		Turf (dry granular)	1 lb. a.i./A	2 lb./1000 sq. ft.	1	—
		Ornamentals (nursery)	0.25–0.5 lb. a.i./100 gal.	100–400 gal./A, runoff	1	—
		Ornamentals (landscape)	0.25–0.5 lb. a.i./100 gal.	Runoff	1	—
Turf	7	Liquid	1 lb. a.i./A	2–4 gal./1000 sq. ft.	1–2	—
		Dry granular	1 lb. a.i./A	2 lb./1000 sq. ft.	1–2	—
Indoor	5	Cockroaches/ants	0.25–0.5%	0.25–0.5 gal./structure	1–4	—
		Fleas	0.25–0.5%	0.5–0.75 gal./structure	1–2	—
		Pet collars	Cats 3%, dogs 8%	—	1–2	—
Perimeter	3	Building foundations (2–3 ft)	0.2–0.5%	1–2 gal./structure	1–4	—
		Around structures (6–10 ft)	0.03–0.06%	10–15 gal./structure	1–4	—
Ornamental	1	Nursery	0.25–0.5 lb. a.i./100 gal.	100–400 gal./A, runoff	1–2	—
		Landscape	0.25–0.5 lb. a.i./100 gal.	Runoff	1–2	—

[a] 1993 data.
[b] Retail: chlorpyrifos purchased and applied by resident/owner. All others are professional market segments: chlorpyrifos applied by a trained, certified, or licensed applicator for a fee at the request of the resident/owner.
[c] Use rates are given in pounds per acre to maintain consistency with chlorpyrifos labels.

Identification of Areas of Vulnerability. Characterization of specific geographic areas for vulnerability to exposures of chlorpyrifos in surface water is complicated by the widespread and varied uses across a large range of agricultural and urban uses. Unlike some major agrochemicals such as the herbicide atrazine, which may be restricted primarily to uses in situations in which natural rainfall patterns produce most of the significant runoff events transporting the chemical into surface water bodies (Solomon et al. 1996), chlorpyrifos is applied in significant volumes in all regions of the United States, often at times when irrigation is employed to supplement natural rainfall. This is particularly true in western irrigated agriculture and in many outdoor urban settings in all regions. Therefore, both rainfall and irrigation can contribute significantly to chemical transport in runoff.

To account for this complexity, areas potentially vulnerable to surface water exposures were identified according to a number of criteria. The first indicator was volume of chlorpyrifos sales by geographic region (Fig. 5–Fig. 7). Greater risk was assumed in areas where chlorpyrifos use is greatest, regardless of use pattern or weather conditions.

Information on sales by county (N. Poletika, Dow AgroSciences, personal communication) was used as a surrogate for chlorpyrifos use in local geographic areas. For agricultural formulations, this assumption is believed to be valid because many distributors are involved in selling chlorpyrifos to a relatively localized customer base, and little movement of chlorpyrifos across county lines is expected. Sales reports from September 1991 through August 1994 were

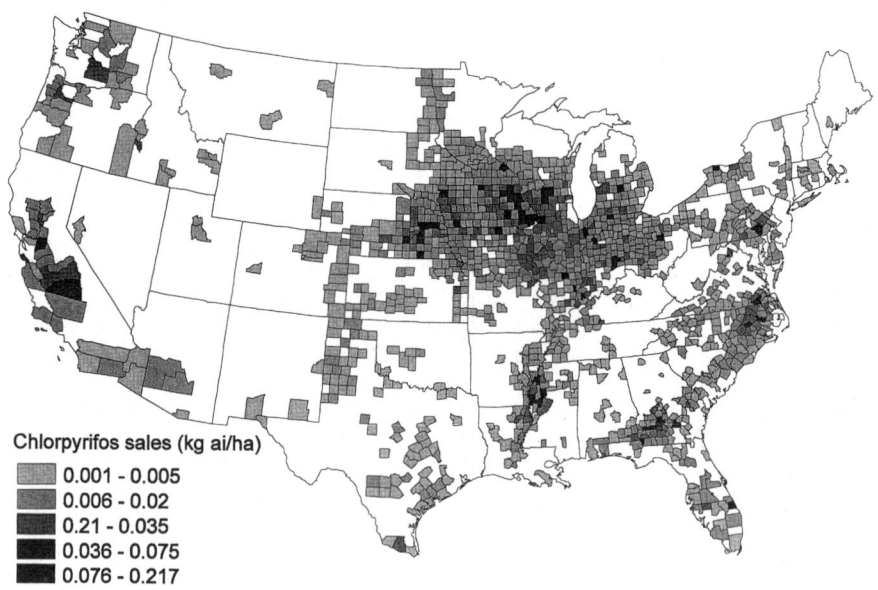

Fig. 5. Agricultural use of chlorpyrifos normalized by county area.

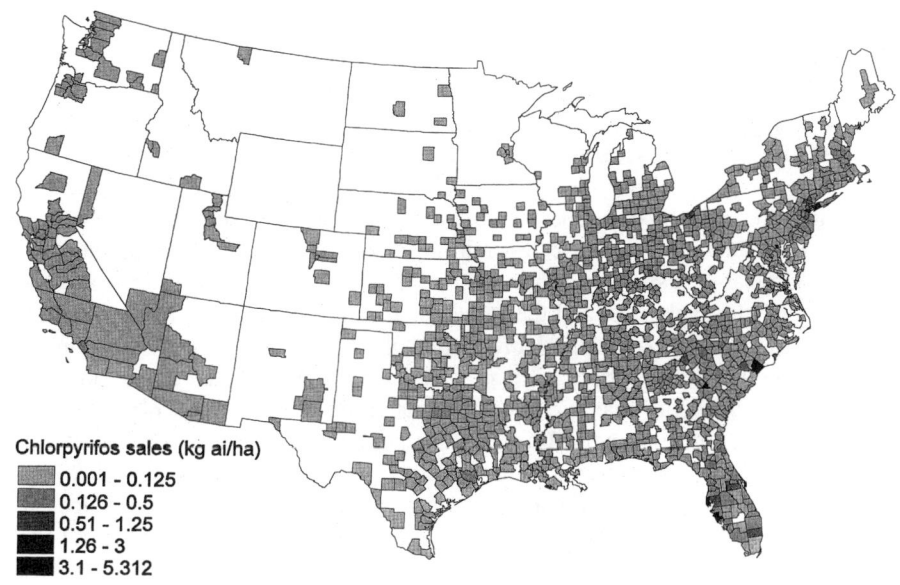

Fig. 6. Urban use of chlorpyrifos normalized for county area.

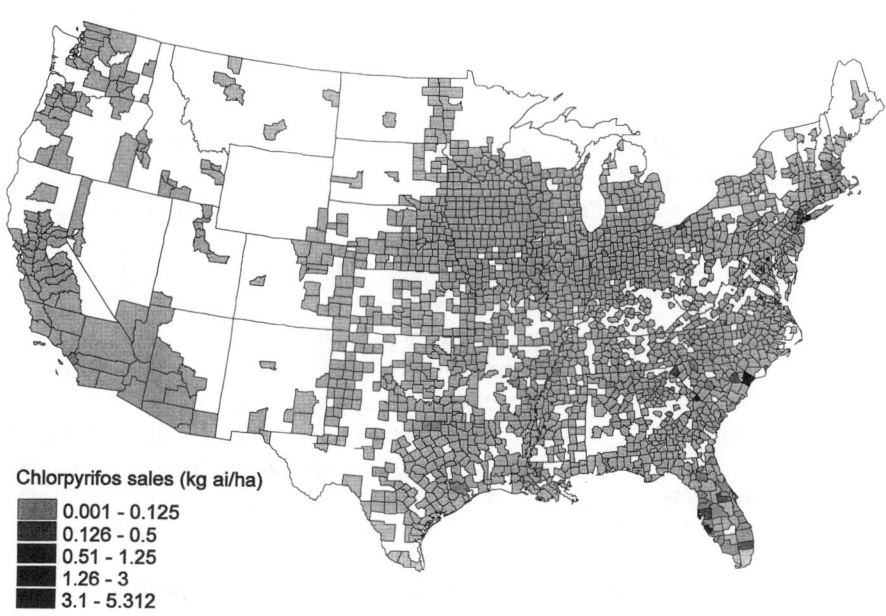

Fig. 7. Combined agricultural and urban use of chlorpyrifos normalized by county area.

summed across all agricultural formulations and averaged over reporting years to obtain mean annual county sales of active ingredient in kilograms (kg). This value was then area-normalized by dividing the mean annual sales volume by the county area in hectares (ha), assuming that every hectare in each county possessed agricultural capability and could receive a chlorpyrifos application (Fig. 5).

Determining the geographic pattern of urban uses is complicated by several factors including the existence of multiple chlorpyrifos formulations, sales accounts with large distributors that receive chlorpyrifos at a single location and redistribute to franchisees, and the existence of third party formulators that market their own products through large chains of retail outlets. Urban sales reports were manipulated to improve their accuracy in predicting actual geographic use areas in the following manner. Sales reports from 1996 were used to approximate sales for the same period covered in the agricultural sales reports (1991–1994). This was done for two reasons. First, 1996 is the earliest year in which tracking of all active ingredient sales was recorded by Dow AgroSciences in sufficient detail to permit this type of analysis. Second, because the rate of use in urban situations is relatively stable, the 1996 data were judged to be representative of sales in prior years for this decade. Active ingredient sales in kilograms (kg) were summed across all formulations of Dow AgroSciences chlorpyrifos and mapped by county. Counties for which unusually great sales volumes were identified were scrutinized for the presence of large distributors that marketed chlorpyrifos either regionally or nationally. These counties were then compared to their nearest neighbors of comparable population and area to determine a representative chlorpyrifos volume for local consumption. The local consumption volume was subtracted from the total delivery to the county, and the remainder was redistributed across the appropriate region by county, apportioned by the fraction of total U.S. sales reported for each county (after correcting for any geographic bias). It was assumed that the geographic distribution of Dow AgroSciences chlorpyrifos formulations would be similar to the distribution and sales of third-party formulations for which the only information available was total sales of active ingredient to all formulators. This total third-party-formulated sales volume was then redistributed nationally, again scaled by the fraction of total U.S. sales of Dow AgroSciences formulations in each county (Fig. 6).

Although there are differences between agricultural and urban use patterns of chlorpyrifos and the specific application sites located in rural and urban settings, it is also true that, after treatments are applied, transport events moving chlorpyrifos offsite into surface water drainage systems can combine chemical residues from disparate sources. Therefore, the sums of the area-averaged agricultural and urban county sales were computed and mapped to provide a picture of the total potential loading of surface water systems from all chlorpyrifos use (Fig. 7).

For use patterns associated with rain-fed agriculture, a vulnerability map based on edge-of-field runoff simulation modeling was employed to locate regions where weather and soil characteristics suggested the possibility of greater risk (Fig. 8). The map is derived from the raw output of modeling described in Havens et al. (1998), where runoff was simulated with the GLEAMS runoff model on a supercomputer. Thousands of simulations were conducted for the contiguous 48 U.S. in areas supporting rain-fed agriculture. Thirty years of simulated weather data were generated by the CLIGEN (Nicks 1989) weather simulator, using parameters from weather stations selected by the geographic intersection of state boundary and Land Resource Region. This simulated weather was input into GLEAMS to simulate runoff from a generic fallow field for each soil type in a regional soil association. The 30-yr average output for each soil type was area-weighted to generate predicted values for runoff water yield and eroded sediment at the association level (Havens et al. 1995).

The sum of relative risk for runoff (water yield) and erosion (sediment yield) by soil association was used as a general indicator of the most sensitive regions. Relative risk was computed by normalizing all water and sediment yield predictions by the most extreme event simulated and then multiplying by 100 to obtain a percentage. The two indices were summed, which produced a maximum combined index value greater than 100% (there were sites where relative water yield and sediment yield index values were both greater than 50%). When the range of the combined index was plotted by soil association, it was found that the extreme events in northwestern Washington State so dominated the set of normalized values that no reasonable classification scheme would produce a map showing useful gradations in vulnerability. Therefore, the data were log-transformed ($\log_{10}$ [index value + 1]), using the constant 1.0 to eliminate zeroes in

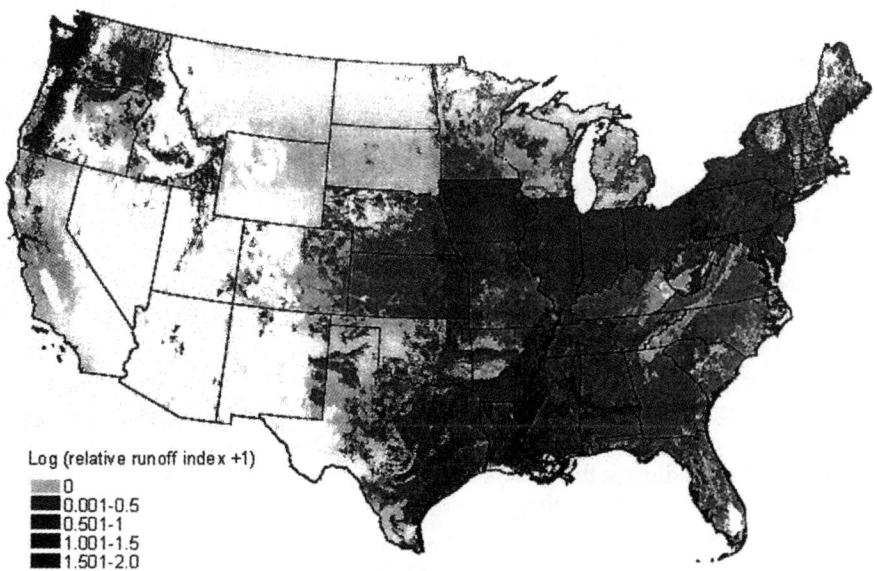

Fig. 8. Map showing generalized potential for chlorpyrifos exposures in surface waters from natural runoff.

the data set. The log-transformed data provided a better visualization of the regional differences in vulnerability in the areas outside northwestern Washington State (Fig. 8). Differences associated with political boundaries reflect cases where climatic data from a single weather station were used to conduct modeling for all local sites simulated within a state (the state boundary intersected with a single Land Resource Region). This map depicts areas where substantial runoff and erosion can occur in fallow ground independent of chemical transport. Runoff water and sediment were combined into a single index to account for the partitioning behavior of chlorpyrifos, which indicates that transport of both dissolved and sediment-sorbed residues can occur (see Table 2). This index of potential risk is conservative for agricultural and turf runoff as the result of the assumption of minimal management practices on fallow ground, where greater runoff and erosion may occur relative to cropped situations. The index may underpredict runoff in urban storm water that is contributed by overland flow across pavement and other hard surfaces limiting infiltration. In most cases, however, the geographically referenced runoff/erosion predictions should provide adequate characterization of generalized potential for exposure to chlorpyrifos. The assumptions made in this analysis do not affect the assessment of the absolute risk as the analysis was used only to identify and rank the major areas of vulnerability.

Specific regional vulnerability to chlorpyrifos exposure in surface water resulting from natural runoff events can be inferred from the combined criteria of chlorpyrifos use volume (Figs. 5–7) and runoff potential (Fig. 8). Relative to agricultural uses, the Midwestern corn belt and Mississippi Delta region were predicted to have the greatest risk of runoff. In these areas there was greater chlorpyrifos use and moderate to great runoff potential. Similar conditions existed in western New York State and southeastern Pennsylvania. In the Southern Atlantic coastal region and in southern Florida, there were areas of greater chlorpyrifos use but comparatively less runoff potential. The remaining greater use region was the Central Valley of California, which is less subject to severe runoff from natural storm events. The other intersecting regions in Fig. 5 and Fig. 8 were intermediate to low in risk of offsite migration.

Urban use volume tended to be concentrated in the most southern tier of states and the eastern seaboard (Fig. 6). Some urban counties in the New York City area, in the Midwest, and in Texas had greater chlorpyrifos use, which coincides with moderate runoff potential (Fig. 8). The reminder of the mapped area exhibited moderate to low relative risk under natural rainfall conditions.

The representation of combined agricultural and urban use in Fig. 7 primarily rescales the classification scheme for the other sales maps. However, the combined sales map indicates that, when both agricultural and urban uses are considered together, the areas of chlorpyrifos use become denser and more filled in (fewer counties showing zero use). This suggests that, where agricultural and urban use occurred together in a watershed, there were more potential sites available to contribute chemical residues to runoff.

Comparable maps for risk of exposure from spray drift were not developed because we judged that drift results from local events independently of geography, is dependent on use pattern, and varies within a region according to the mix of use patterns and water bodies present. Therefore, in this assessment the contribution of spray drift was not assessed, acknowledging that it can be important in transient exposures of aquatic organisms to chlorpyrifos.

B. Environmental Fate and Behavior

Environmental Persistence. Chlorpyrifos (Fig. 9) will degrade by both abiotic and biotic transformation processes in terrestrial and aquatic environments. In soil, water, plants, and animals, the major pathway of abiotic and biotic degradation involves cleavage of the phosphorothioate ester bond (Racke 1993) to form 3,5,6-trichloro-2-pyridinol (TCP. CAS 6515-38-4; Fig. 9). TCP is degraded in the environment via photolysis with an aqueous half-life of about 4 min in surface water at 40 °N latitude (Dilling et al. 1984) and microbial degradation with an average soil half-life of 73 d at 25 °C (Bidlack 1976).

In terrestrial ecosystems, chlorpyrifos rapidly dissipates from plant foliage, with an observed half-life of <1 to 9 d (Racke 1993). Chlorpyrifos dissipates at a moderate rate when incorporated into the soil profile; half-lives of 33–56 d were noted at sites in California, Michigan, and Illinois (Fontaine et al. 1987). However, dissipation from soil surfaces occurs rapidly. Soil surface half-lives of 9–11 d have been noted for fallow soil surfaces and from 7 to 9 d from turf grass surfaces following broadcast spray applications at sites in Indiana and Florida (Racke and Robb 1993).

In aquatic ecosystems, chlorpyrifos is removed from the water column via hydrolysis, biodegradation, sorption to sediment, volatilization, and photodegradation. Hydrolysis half-lives in sterile, distilled water have been reported to range from 16 to 72 d at pH 5–9, while laboratory photolysis half-lives of 30–52 d have been reported (Racke 1993). Degradation half-lives in sediment–water laboratory systems under aerobic and anaerobic conditions have been reported as 22–51 and 39–200 d, respectively (Racke 1993). However, several research-

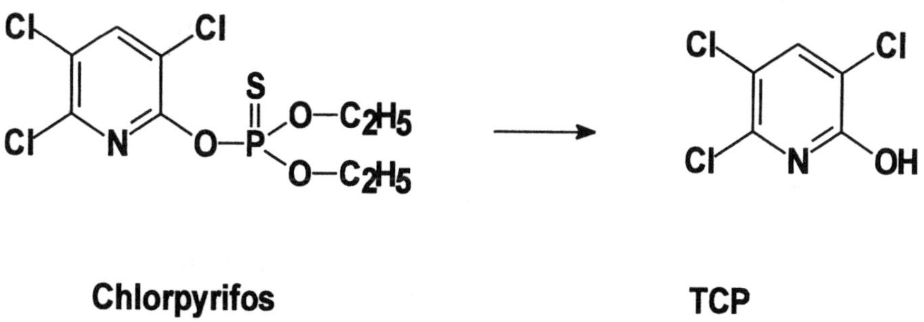

Chlorpyrifos   **TCP**

Fig. 9. The structure of chlorpyrifos and its primary metabolite, TCP (3,5,6-trichloro-2-pyridinol).

ers have examined the behavior of chlorpyrifos in natural water and sediments and generally observed shorter dissipation half-lives than for laboratory studies. The more rapid decrease in concentrations under field conditions is likely caused by additional dissipation and degradation forces such as volatilization and surface- and metal-catalyzed hydrolysis that operate in natural waters and sediments and are not assessed under laboratory conditions, such as reflected in the data in Table 2. Water column half-lives ranging from 0.08 to 5 d and sediment half-lives of 0.8–16.3 d have been reported from a number of field investigations (Racke 1993).

Pathways of Environmental Transformation. Chlorpyrifos can undergo several types of biological and chemical transformations in the environment. The two most important degradation processes are hydrolysis and oxidation (Fig. 10).

The principal hydrolysis product, the trichloropyridinol (Fig. 10), is the single most important transformation product in water as well as soil or other matrices. In addition, it is the metabolite formed in greatest quantities in vivo by most organisms. The hydrolysis can be achieved either chemically or biologically, the latter mostly through the action of phosphatase enzymes. Concomitant with the formation of the trichlopryidinol is the production of diethylphosphorothioate. In an aqueous medium, the hydrolysis of chlorpyrifos is strongly influenced by the pH of the water. Alkaline conditions enhance the rates of hydrolysis and contribute to shorter half-lives in some aquatic media.

The primary oxidation product is chlorpyrifos oxon, in which the thion sulfur

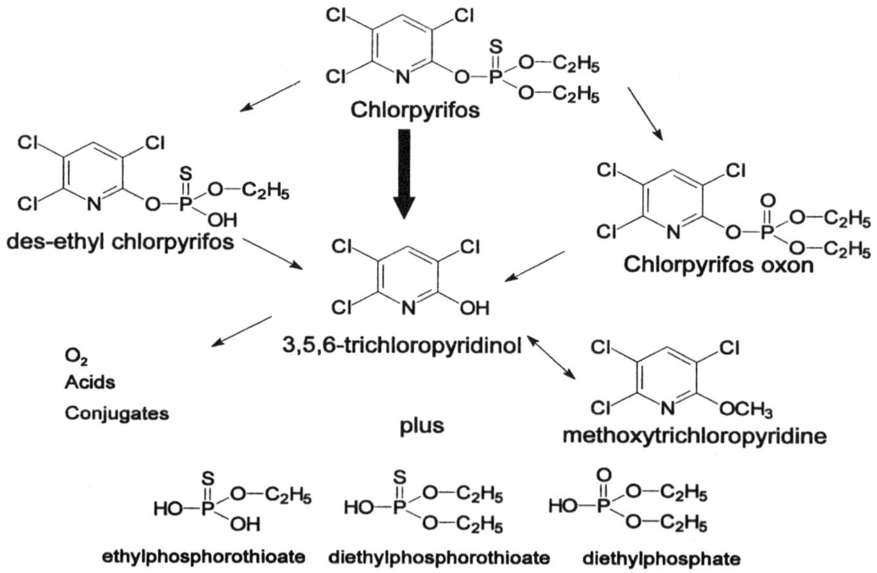

Fig. 10. Environmental transformations of chlorpyrifos (adapted from Racke 1993).

has been oxidized and replaced by an oxygen atom (Fig. 10). The oxon is very susceptible to hydrolysis and is hence quickly degraded further, principally to the trichloropyridinol and diethylphosphate. Another biotransformation that can occur through an oxidative mechanism is the deethylation of the parent molecule or the oxon. The dealkylation of organophosphorus esters can also occur via hydrolysis or through the action of glutathione-S-transferase enzymes. The resultant des-ethyl chlorpyrifos can subsequently undergo the same hydrolysis or oxidation reactions as the parent molecule.

The principal degradation product, the trichloropyridinol, is ultimately degraded to carbon dioxide by microbes and other organisms. It can also be methylated in some organisms to form methoxytrichloropyridine, which can be readily transformed back to the pyridinol form and ultimately conjugated or degraded to carbon dioxide.

C. Measures of Exposure

Estimates of Environmental Concentrations in the EPA Tiered Approach. The hazard quotient method (HQ) of risk analysis (Urban and Cook 1986) compares an expected environmental concentration (EEC) to a measure of hazard to aquatic organisms (for acute effects, typically the LC_{50} for a sensitive species or some benchmark of other adverse effect).

For a typical EPA tiered exposure assessment, various methods are used to estimate EECs. Tier I utilizes a simple model that considers a few basic chemical parameters and application information and is conservative. This tier makes very conservative assumptions and is used only as a one-tailed test to rule out compounds that pose little risk to nontarget organisms. More refined tiers employ more sophisticated numerical simulation models to predict environmental concentrations. The next sections describe calculations of EECs for chlorpyrifos obtained for corn, the typical agricultural field use, in a tiered assessment process.

Tier I. A Tier I exposure assessment for a T-band application of 1 kg/ha ($\approx$1.2 lb a.i./acre) in field corn was performed using the EPA GENEEC simulation model (Parker and Rieder 1995). The simple GENEEC model assumes that 100% of the surface runoff from a 10-ha treated field enters an adjacent 1-ha farm pond with a depth of 2 m. Peak (maximum point estimate) and 96-hr average EECs resulting from this assessment were 2560 and 2160 ng/L, respectively.

Tier II. A Tier II exposure assessment for at-plant use in field corn has been conducted (USEPA 1996a). This tier represents a refined assessment of exposure, but is still conservative and is used as a screening method. Two sites for modeling were selected to represent corn culture in the U.S., one a typical "medium exposure" site in Iowa and the other a "high exposure" site in Mississippi. Numerical modeling using the EPA models PRZM1 to simulate surface runoff and EXAMS to simulate fate in the aquatic environment was performed. It has been assumed that there would be 100% runoff from a 10-ha treated field to an adjacent 1-ha farm pond with a depth of 2 m. Thirty-six years of actual weather data from National Weather Service stations in the simulated regions were used in the runoff simulations to obtain a distribution of EECs in the farm pond at the two sites. Results were reported for a "typical worst case," defined as the 10-yr return period for each simulation, between the third and fourth years of greatest chemical runoff predictions in the 36-yr simulated period.

For a banded application rate of 1.16 kg/ha (1.3 lb a.i./acre), typical and "worst-case" weather-driven EECs were obtained (Table 5). In contrast to Tier I, this assessment introduced some aspects of probability into the analysis by selecting a 90th centile EEC and by considering two sites to represent median and reasonable high exposure site vulnerabilities (R. Parker, USEPA, personal communication).

Tier III. The intent of the Tier III assessment was to expand runoff and pond fate simulation modeling over a relatively great number of representative sites in a region to generate a probabilistic description of EECs that more accurately reflect the influence of weather and site characteristics found across the region. Because the EPA Tier III assessment method is not yet implemented, a system of environmental fate models was applied to generate frequency distributions of predicted EECs for regions in which major chlorpyrifos-treated crops are grown. A similar assessment for field corn grown in the Midwestern U.S. corn belt was reported by Havens et al. (1998).

A geographically based, probabilistic modeling system was utilized to perform the analyses. This system combined geographic information, database management, and numerical simulations to define regions of interest, determine simulation scenarios describing the regions, realize the simulations, and geographically reference the results. Quantities of chlorpyrifos sold in a region were used as a surrogate estimate of quantities used. This information was combined with various sources of land use information to define the regions of interest for each crop. Within the regions of interest, all areas mapped to soil associations supporting commercial production of the simulated crop were included in the assessment. Field-scale simulations of runoff were performed by the USDA-ARS model GLEAMS, using simulated weather patterns selected to describe

Table 5. Tier II expected environmental concentrations (EECs).

Site	Time	EEC (ng/L)
Iowa	Peak	2,800
	96-hr	2,300
Mississippi	Peak	13,000
	96-hr	10,200

the typical frequency of "worst-case" storm events. The storms in the simulated weather patterns were applied to each soil association within the area of influence of the weather station providing the runoff modeling parameters. Weather data were obtained from the USDA-ARS climate-generating program CLIGEN (Nicks 1989). Simulations assumed complete drainage of runoff from a 10-ha treated field into an adjacent 1-ha farm pond 2 m in depth. Fate of runoff inputs into the pond was simulated with the EPA model EXAMS II.

The distribution of predicted peak EECs up to the 90th centile of corn acreage was calculated as concentration-acreage classes in field corn treated with a T-band application of 1 kg/ha (1.2 lb a.i./acre) in conventional tillage (Table 6). These EECs can be interpreted as "typical worst-case" hazards from extreme storms driving runoff events across a distribution of site vulnerabilities that were defined by local factors such as soil type, slope, and weather conditions.

In the Tier III assessment, the peak Iowa EEC of 2,800 ng/L calculated in Tier II was predicted to occur across approximately 15% of corn belt acreage. Based on the Tier III analysis, the peak Mississippi EEC of 13,000 ng/L predicted in Tier II is unlikely to be observed in the corn belt (<10% of the acreage). The edge-of-field, regional-scale runoff modeling in the Tier III exposure assessment most probably overpredicts EECs, as do the less sophisticated Tier I and Tier II assessments. Factors that may contribute to overprediction of the EEC include conservative, typical worst-case assumptions for weather, field drainage, and possible mitigation by buffer zones at field borders. There is also minimal dilution and dispersion of pulsed inputs into the constant-volume, static farm pond, and this environment is not representative of the large reservoirs or flowing water systems that may be considered more important ecosystems requiring protection. The capability of existing environmental fate models to give accurate edge-of-field predictions is also in question (Solomon et al. 1996). Thus, it was concluded that additional refinement of the exposure assessment was necessary.

Table 6. Tier III peak EECs for midwestern U.S. field corn.

Percent of simulated area[a]	Cumulative percent of simulated area	Peak EEC (ng/L)
25.2	25.2	1000
21.2	46.4	2000
14.9	61.3	3000
8.4	69.7	4000
6.8	76.5	5000
5	81.5	6000
3.6	85.1	7000
3.2	88.3	8000
2.1	90.4	9000

[a]Total number of hectares (ha) simulated, 2.2 million ha (249,740,324 acres).

Tier IV. No specific approach to conducting a Tier IV exposure assessment has been adopted by EPA. It has been suggested that the appropriate refinement is to consider landscape factors in a watershed rather than at the edge of the treated field (SETAC 1994). The assessment tool needed for this refinement is landscape modeling that accounts for the location of treated areas relative to surface waters and actual percentages of land treated in a landscape unit. Presently, there are no simulation models that can reliably estimate pesticide EECs in surface waters on a watershed scale. The alternative assessment tool to modeling is surface water monitoring, at least for existing products with widespread commercial use. When compounds are in current use, monitoring data can be used to provide estimates of actual environmental concentrations, and when information on use patterns is combined with watershed vulnerability estimates, these concentrations can be interpreted in a meaningful way for risk analysis. A relatively great amount of monitoring data for chlorpyrifos was available, and the detailed exposure assessment that follows is based on empirical measurements of environmental concentrations.

Measured Environmental Concentrations in Freshwater. Recent data on concentrations of chlorpyrifos in surface waters were available from several sources. The most extensive data set found was for the Midwestern U.S. corn belt. This data set was generated by the Heidelberg College Water Quality Laboratory as part of an ongoing monitoring program of rivers and tributaries draining into Lake Erie. A portion of this data set has been previously reported (Richards and Baker 1993). Data from seven sites for the period 1983 through mid-1995 were obtained (R.P. Richards, personal communication, June 1996). The locations of the drainage basins sampled by each monitoring site, periods of record, and numbers of observations are summarized (Fig. 11 and Table 7). In general, the sampling frequency was three samples per day from April 15 to August 15 (pesticide runoff season) and two samples per month for the remainder of the year. During the pesticide runoff season, all samples related to a runoff event were analyzed for chemical residues, while only two samples per week were analyzed during low-flow conditions. Unfiltered samples were prepared by liquid–liquid extraction and analyzed either on a dual-column gas chro-

Table 7. Sampling sites for the Lake Erie drainage basin

Stream	Area (km^2)	Percent cropped	Dates of operation	Number of days sampled
Lost Creek, OH	11	83	1983–1993	515
Rock Creek, OH	88	81	1983–1995	837
Honey Creek, OH	390	83	1983–1995	916
Huron River, OH	930	73	1988–1991	191
River Raisin, MI	2,700	67	1983–1995	829
Sandusky River, OH	3,200	80	1983–1995	865
Maumee River, OH	16,000	76	1983–1995	829

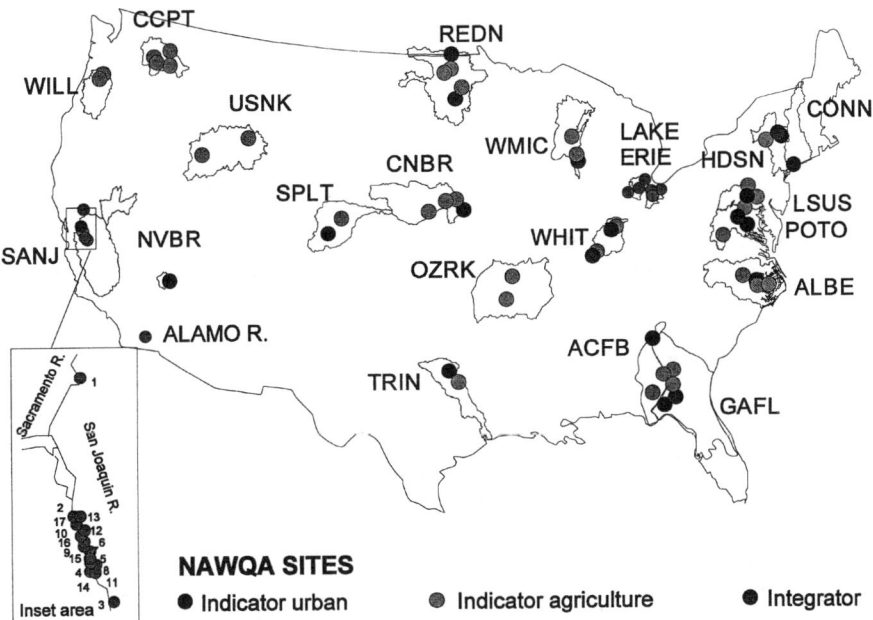

Fig. 11. Location of watersheds and sampling sites where chlorpyrifos concentration data were obtained. Site numbers in the inset correspond to those in Table 8. NAWQA (National Water Quality Assessment Program) codes are listed in Table 9

matograph (GC) using nitrogen-phosphorus detection or by GC with mass selective detection. The least sensitive reported limit of quantification (LOQ) reported for chlorpyrifos was 20 ng/L (Richards and Baker 1993).

In California, where chlorpyrifos use is widespread, two government agencies have monitored for pesticides in surface waters of the Central Valley agricultural area for various periods of time. The largest data set comes from a 1991–1994 U.S. Geological Survey (USGS) study on the Sacramento River at Sacramento and on the San Joaquin River at Vernalis (MacCoy et al. 1995). These sites were selected because they are downstream from most tributaries in the Sacramento and San Joaquin valleys, respectively (Fig. 11), and therefore the concentration data can be used with discharge data to calculate loadings entering the San Francisco Bay Estuary. Concentrations were measured in filtered water samples using solid-phase extraction and gas chromatography-mass spectrometry (GC-MS). The dates of collection, numbers of samples, and method detection limits are summarized in Table 8.

Monitoring conducted by the California Department of Pesticide Regulation (DPR) generated data from 15 other locations on the San Joaquin River upstream of Vernalis and on tributaries and constructed drains terminating at the river (Ross 1992a, b, 1993a–c) (Fig. 11, Table 8). Sample handling procedures and analytical methods were the same as those used by the USGS.

Table 8. Chlorpyrifos sampling locations in the U.S. Geological Survey (USGS) and state monitoring programs in California.

Site no.	Site[a]	Agency	Dates of operation	Number of days sampled	Method detection limit (ng/L)
1	Sacramento R. at Sacramento	USGS	5/91–4/94	471	28–44
2	San Joaquin R. at Vernalis	USGS	1/91–4/94	1,122[b]	28–35
3	Salt Slough	DPR	3/92–6/92	10	50
4	Orestimba Creek	DPR	2/91–6/92	19	50
5	Spanish Grant Combined Drain	DPR	3/91–6/92	19	50
6	TID No. 3 Drain	DPR	3/91–6/92	17	50
7	TID No. 5 Drain	DPR	3/91–6/92	22	50
8	TID No. 6 Drain	DPR	5/91–6/92	17	50
9	Del Puerto Creek	DPR	3/91–6/92	22	50
10	Ingram/Hospital Creeks	DPR	3/91–6/92	24	50
11	Merced River	DPR	4/91–2/93	17	50
12	Tuolumne River	DPR	1/92–6/92	11	50
13	Stanislaus River	DPR	4/91–6/92	12	50
14	San Joaquin R. at Hills Ferry Road	DPR	5/92–9/92	37	50
15	San Joaquin R. at Patterson	DPR	7/91–9/92	26	50
16	San Joaquin R. at Laird Park	DPR	3/91–2/93	122	50
17	San Joaquin R. at Airport Road	DPR	5/91–6/92	14	50
	Alamo River at Harris St. Bridge	SWRCB	9/93–2/94	15	50–100

TID, Turlock Irrigation Ditch, Ca; DPR, Department of Pesticide Regulation; SWRCB, State Water Resource Control Board.
[a]Sites with at least 10 observations.
[b]Samples collected in 2 or more consecutive days usually were combined for analysis, except during periods of rainfall.

The State Water Resources Control Board (SWRCB) initiated a study with the University of California at Davis Aquatic Toxicology Laboratory to monitor agricultural drains of the Colorado River Basin in the Imperial Valley (DiGiorgio et al. 1995) (Fig. 11, Table 8). Filtered samples collected from the Alamo River were analyzed by DPR and two contract laboratories using similar gas chromatography methods.

Concentrations of chlorpyrifos in surface waters measured in the first group of 20 study units of the USGS National Water Quality Assessment Program (NAWQA) were obtained (R. Gilliom, personal communication, January 1997). As indicated in the NAWQA implementation plan (Leahy et al. 1990), this program intends to provide a nationally consistent description of current water quality conditions for a large portion of U.S. water resources. An important specific water quality concern to be addressed by the NAWQA program is the occurrence and concentration of pesticides in surface waters and their relation to aquatic health criteria. Rivers and streams in study units established in representative hydrologic basins across the U.S. were sampled from 1992 through 1995 and analyzed at the National Water Quality Laboratory at Denver (Zaugg et al. 1995). Two types of sampling sites were typically selected in each study unit: indicator and integrator (Table 9). Indicator sites represent a small drainage basin, of tens to hundreds of kilometers squared (km^2), with relatively homogeneous land use: agricultural (indicator-agriculture) or urban (indicator-urban). The integrator sites sample a large drainage basin, hundreds to thousands of km^2 with mixed land use (Fig. 11). Concentrations were measured in filtered water samples using solid-phase extraction and GC-MS. The method detection limit for chlorpyrifos was set at 4 ng/L (R. Gilliom, personal communication, January 1997).

Chlorpyrifos Residues in Fish. A 1992 study of residues of chemicals in fish in U.S. rivers by the USEPA (USEPA 1992b) found that less than 30% of the fish sampled contained detectable concentrations of chlorpyrifos. For these fish, many samples were associated with agricultural production, and the greatest concentrations measured in fish were from sites already identified as areas of greatest potential exposure to chlorpyrifos, based on concentrations in water. Because of a lack of extensive knowledge of chlorpyrifos body dose in respect to toxicity, these data are not readily usable for risk assessment purposes. However, they do serve as a confirmation that the sites selected for water column sampling were consistent with sites where fish were exposed to chlorpyrifos.

Measured Concentrations in Saltwater. Concentrations of chlorpyrifos have been measured in the waters of Chesapeake Bay on four occasions in 1993 and at eight different sites (McConnell et al. 1997). Samples (50 L) were extracted on XAD-2 resin and analyzed by GC-MS with a method detection limit of 0.045 ng/L.

Analysis of Freshwater Monitoring Data. All observations were included in the distributional characterization of exposure, even those where chlorpyrifos was not detected. In all cases, only data sets with 10 or more observations were included in the analysis.

Analysis of Data. Uncensoring methods were considered as a procedure for considering nondetects; however, it was concluded that these methods would yield values of little relevance to the risk characterization (small concentrations in relation to the toxicity measures). Therefore, concentrations reported as nondetects (ND, below the limits of detection [LOD] of the analytical method) were

Table 9. National Water Quality Assessment Program (NAWQA) Group 1 Study Unit sampling locations.

Study unit name	Sampling site name[a]	Code	Type	Dates of operation	No. of days sampled
Apalachicola-Chattahoochee-Flint R. basin	Aycocks Creek near Boykin GA	acfb-aycocks	incator-agriculture	3/93–7/94	34
	Lime Creek (County road) near Cobb GA	acfb-lime	indicator-agriculture	3/93–4/95	54
	Sope Creek (S Roswell RD) near Marietta GA	acfb-sope	indicator-urban	3/93–2/95	52
Albemarle-Pamlico drainage	Albemarle Canal near Swindell NC	albe-albe	indicator-agriculture	3/93–8/94	34
	Devils Cradle Creek at Sr 1412 near Alert NC	albe-devils	indicator-agriculture	3/93–8/94	27
	Pete Mitchell Swamp at Sr 1409 near Penny Hill NC	albe-pete	indicator-agriculture	3/93–8/94	35
	Tar River at Tarboro NC	albe-tar	int	3/93–8/94	20
Central Columbia plateau	Crab Creek at Marcellus Road near Ritzville WA	ccpt-crab.m	indicator-agriculture	4/93–1/95	21
	Crab Creek Lateral Ab Royal Lake near Othello WA	ccpt.crab.rl	indicator-agriculture	3/93–2/95	33
	EL 68 D Wasteway near Othello WA	ccpt-el68	indicator-agriculture	4/93–2/95	31
	Palouse River at Hooper WA	ccpt-palouse	indicator-agriculture	3/93–2/95	40
Central Nebraska basins	Maple Creek near Nickerson NE	cnbr-maple	indicator-agriculture	5/92–3/94	59
	Platte River at Louisville NE	cnbr-platte	integrator	5/92–3/94	59
	Prairie Creek near Ovina NE	cnbr-prairie	indicator-agriculture	4/93–3/94	16
	Shell Creek near Columbus NE	cnbr-shell	indicator-agriculture	4/93–3/94	15

Connecticut	Norwalk River at Winnipauk CT	conn-norwalk	indicator-urban	3/93–8/94	51
Georgia-Florida coastal plain	Lafayette Creek Miccosukee Rd (No. 28) TLH FL	gafl-lafayette	indicator-urban	3/93–2/95	41
	Little River at Upper Tyty Rd Near Tifton GA	gafl-little	indicator-agriculture	3/93–4/95	37
	Tucsawhatchee Creek (GA HWY 27) near Hawkinsville GA	gafl-tucsa	indicator-agriculture	3/93–5/95	53
	Withlacoochee River at US 84 near Quitman GA	gafl-withla	integrator	3/93–5/95	37
Hudson River basin	Canajoharie Creek Near Canajoharie NY	hdsn-canaj	indicator-agriculture	3/94–4/95	21
	Lisha Kill northwest of Niskayuna NY	hdsn-lisha	indicator-urban	3/94–4/95	22
	Mohawk River at Cohoes NY	hdsn-moh	integrator	3/94–4/95	20
Lower Susquehanna River basin	Cedar Run at Eberlys Mill PA	lsus-cedar	indicator-urban	3/93–5/95	62
	East Mahantango Creek at Klingerstown PA	lsus-eastm	indicator-agriculture	3/93–9/94	44
	Mill Creek at Eshelman Mill Road near Lyndon PA	lsus-mill	indicator-agriculture	3/93–9/94	43
Nevada basin and range	Las Vegas wash below Flamingo wash Confluence near Las Vegas NA	nvbr-lasvegas	indicator-urban	5/93–3/95	40
Ozark plateaus	Dousinbury Creek on JJ near Wall Street MO	ozrk-dous	indicator-agriculture	2/94–12/94	22
	Yocum Creek near Oak Grove AR	ozrk-yocum	indicator-agriculture	2/94–11/94	23
Potomac River basin	Accotink Creek near Annandale VA	poto-acco	indicator-urban	3/94–4/95	30

(Continued)

Table 9. Continued.

Study unit name	Sampling site name[a]	Code	Type	Dates of operation	No. of days sampled
	Monocacy River at Bridgeport MD	poto-mono	indicator-agriculture	3/94–5/95	36
	Muddy Creek at Mount Clinton VA	poto-muddy	indicator-agriculture	3/93–5/95	34
	Shenandoah River at Millville WV	poto-shenan	integrator	3/93–7/94	18
Red River of the North basin	Red River of The North above Fargo ND	redn-redn	integrator	4/94–3/95	10
	Red River of The North at EMERSON	redn-rr.em	integrator	4/93–5/94	22
	Snake River above Alvarado MN	redn-snake	indicator-agriculture	5/93–9/94	27
	Turtle River at Turtle River State Park near Arvilla ND	redn-turtle	indicator-agriculture	3/93–9/94	30
	Wild Rice River at Twin Valley MN	redn-wildr	indicator-agriculture	3/93–8/94	27
San Joaquin-Tulare basins	Merced River near Stevinson CA	sanj-merced	indicator-agriculture	1/93–6/94	50
	Orestimba Creek at River Rd near Crows Landing CA	sanj-orest	indicator-agriculture	4/92–3/95	85
	Salt Slough at Hwy 165 Near Stevinson CA	sanj-salt	indicator-agriculture	1/93–6/94	26
	San Joaquin River near Vernalis CA	sanj-sj	integrator	4/92–3/95	73
South Platte basin	Cherry Creek at Denver CO	splt-cherry	indicator-urban	3/93–11/94	36
	Lonetree Creek near Greeley CO	splt-lone	indicator-agriculture	4/93–8/94	35
Trinity River basin	Chambers Creek near Rice TX	trin-chamb	indicator-agriculture	6/93–4/95	28

Basin	Site	Code	Type	Dates	N
	Rush Creek at Woodland Park Blvd. Arlington TX	trin-rush	indicator-urban	3/93–5/94	22
Upper Snake River basin	Rock Creek above Hwy 30/93 crossing at Twin Falls ID	usnk-rock	indicator-agriculture	4/93–5/95	41
	Teton River near St Anthony ID	usnk-teton	indicator-agriculture	5/93–9/94	29
White River basin	Kessinger Ditch near Monroe City IN	whit-kess	indicator-agriculture	3/93–4/95	45
	Little Buck Creek near Indianapolis IN	whit-little	indicator-urban	5/92–3/95	96
	Sugar Creek at County Rd 400 S at New Palestine IN	whit-sugar	indicator-agriculture	5/92–3/95	72
	White River at Hazleton IN	whit-white	integrator	5/92–4/95	72
Willamette basin	Pudding River at Aurora OR	will-pudding	indicator-agriculture	4/93–5/95	28
	Zollner Creek near Mt. Angel OR	will-zollner	indicator-agriculture	4/93–5/95	29
	Fanno Creek at Durham OR	will-fanno	indicator-urban	3/93–5/95	28
Western Lake Michigan drainage	Duck Creek at Seminary Road near Oneida WI	wmic-duck	indicator-agriculture	3/93–6/95	42
	Milwaukee River at Milwaukee WI	wmic-milw	integrator	4/93–9/94	34
	North Branch Milwaukee River near Random Lake WI	wmic-nbmilw	indicator-agriculture	3/93–9/94	36

^aSites with at least 10 observations.

assigned a dummy value for the purposes of the distributional analysis. This procedure is different from the (USEPA) practice of assigning a value of one-half the LOD to the non-detects and that would likely overestimate exposure (Fig. 12). It is also unlikely that all the NDs would be exactly equal to a single fraction of the LOD. Therefore, a more realistic estimator was used in the probabilistic procedure, and the assumption was made that the NDs are distributed from the LOD to true zero, in a continuation of the distribution of the detected concentrations. Thus, in the estimation procedure used, all NDs contribute to the description of the exposure profile without introducing bias to the distribution of the measured concentrations.

Because of differences in sampling procedures, data sets were analyzed using different methods. Some of the data sets were relatively small and some consisted of samples taken primarily during the use season. Data in this category include the NAWQA set and the DPR and SWRCB California sets. This biases the data toward the period of greatest use but, as the crop growing season is also the most productive period in the rivers and streams, the use of seasonally biased sampling data was judged to be both conservative and appropriate. Be-

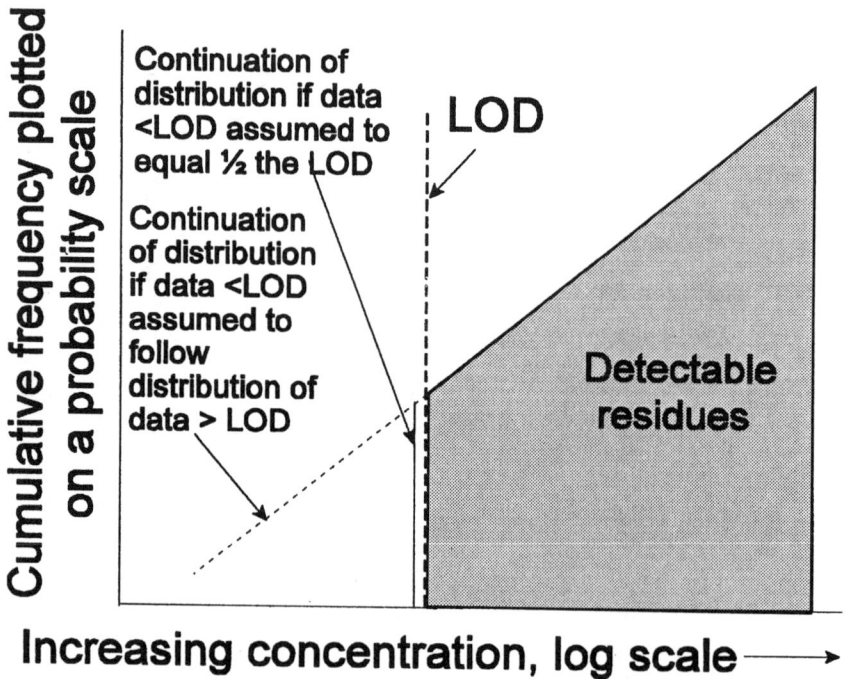

Fig. 12. Illustration of treatment of measurement data lower than the limits of detection (LOD).

cause that is the period when maximum concentrations are expected, such an analysis should provide maximum environmental protection. The USGS California study on the Sacramento and San Joaquin Rivers is the only monitoring program that made closely spaced year-round observations.

Monitoring of the Lake Erie drainage basins conducted by Heidelberg College provided many observations over multiple years with sampling intervals ranging from hours (multiple samples/d collected during runoff events) to months (winter season). This data set was the most comprehensive of all the data sets currently available and is representative of the area of greatest chlorpyrifos use, at-plant applications in Midwestern U.S. field corn. For these reasons, data were selected from this set to conduct a detailed analysis of exposure patterns. While this is not representative of the conditions prevalent throughout North America, it does represent one of the areas believed to be at greatest risk of exposure. To accomplish this analysis, a concentration value was estimated for every day falling within the period of observation. On days for which multiple observations were collected, a daily arithmetic mean value was calculated. Missing observations were estimated by linear interpolation between dates with reported concentrations. The linear interpolation procedure used is conservative when it generates many relatively great concentrations that probably would not have been observed and liberal when it produces a long string of zero values. The greatest uncertainty in the interpolated values lies within periods of months both early and late in the year when, in certain years, no samples were collected. Usually, but not always, a great number of samples in which chlorpyrifos was nondetectable resulted from interpolation between the last observation of the previous reporting period and the first observation of the next period of sampling. For the at-plant corn use pattern, this uncertainty is relatively unimportant as it can be assumed that chlorpyrifos dissipation in treated soil would be rapid, thus limiting the potential for new chemical inputs to enter streams in runoff events occurring after harvest and before new applications were made the following spring.

Moving averages were calculated for the Lake Erie data sets using a time interval of 2 d. A 2-d wide window was moved through time in steps of 1 d. Concentrations in the window were added together and divided by 2, thus giving a time-weighted mean concentration. For all these data sets, distributions of measured data were plotted on a probability or linear percentage scale using a ranked set of concentrations. Plotting positions were calculated from Eq. 1.

$$\frac{100 \times i}{n+1} \tag{1}$$

where $i = i^{th}$ observation of a total of n observations (Parkhurst et al. 1995) and were expressed as cumulative percentages. The concentration values used in this analysis were initially assumed to be log-normally distributed. This distribution has been observed in analyses of exposure to other pesticides (Solomon et al.

1996; Solomon and Chappel 1997). When plotted on logarithmic concentration-probability axes, log-normally distributed data approximate a straight line. Thus, the application of log-probability transformations allows the use of linear regression techniques to obtain a function from which the probability of obtaining an arbitrary concentration within the range of observed values can be calculated (Fig. 13). However, as the data analysis proceeded, some of the distributions were observed to more closely approximate a normal distribution and were plotted on a linear concentration scale versus cumulative probability and not log-transformed (Fig. 13). Some of the ranked data sets possessed sigmoidal distributions. These were plotted on log concentration–cumulative percentage axes and fit to a four-parameter logistic function (Eq. 2).

$$y = d + \frac{(a-d)}{1 + \left(\frac{x}{c}\right)^b} \quad (2)$$

where:
a = the asymptotic maximum
b = the slope parameter
c = the value at the inflection point
d = the asymptotic minimum

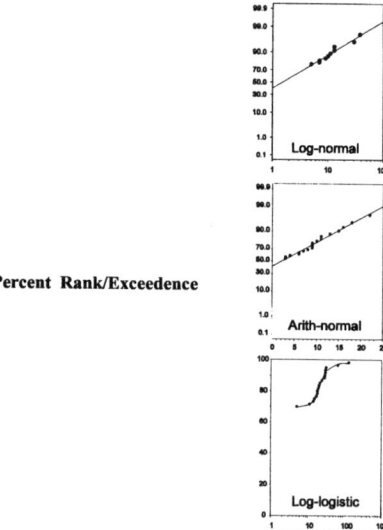

Percent Rank/Exceedence

Fig. 13. Example of a concentration data set showing log-normal, arithmetic-normal, and log-logistic distributions for the NAWQA data set.

The fitted function was then used to provide the best possible estimate of exceedences of given concentrations. In the case of sigmoidal distributions, a percentage scale was used rather than a probability scale because the nonlinear plot violates the assumption of log-normality on a probability scale (Fig. 13). Data were plotted, transformed, and linear or nonlinear regressions performed with a graphics package (SigmaPlot 1997).

Results. Regression coefficients were determined by nonlinear least squares fitted regressions (Eq. 2) for the Lake Erie drainage basin (Table 10). The regression coefficients were used to estimate 90th centiles for 48-hr time-weighted mean concentration (Table 11). The greatest 90th centile concentration, 37 ng/L, was measured in the Huron River, OH, and the least, 3 ng/L, was observed in Lost Creek, OH. There was no apparent relationship between watershed size and magnitude of concentration. Yearly annual maximum measured concentrations of chlorpyrifos in the Lake Erie drainage basin data set were identified (Table 12). Because of the small size of the annual maximum data set, 90th centiles of annual maxima were not calculated. These maxima represent daily concentrations. Inspection of the raw data revealed that these concentrations were not bracketed by similar concentrations and that, within the constraints of the sampling frequency, these peaks occurred in relative isolation (Figure 14).

The 90th centile chlorpyrifos concentrations at California locations were estimated from log-normal linear regression coefficients. These concentrations tended to be greater relative to the Lake Erie drainage basin, at least in the smaller tributaries to the San Joaquin River and in the Alamo River (Table 13). Concentrations were less in the larger tributaries of the San Joaquin River and in the river itself. At the downstream sites at Vernalis and Sacramento (Sacramento River), little chlorpyrifos was detected. The 90th centile concentrations at these locations ranged from ND (0) to 672 ng/L. The greatest concentrations were found in agricultural and combined agricultural/municipal drains. During the monitoring period for the Alamo River there was no measurable rainfall in the Imperial Valley, and all flow downstream of the water delivery canal was from irrigation return water.

Examples of log-normal, normal, and sigmoidal distributions were observed for locations in the NAWQA data set (Table 14). Ninetieth centile concentrations varied from ND (0) to a maximum of 95 ng/L. When the data were pooled by type of site (indicator-agriculture, indicator-urban, integrator), three trends were observed (Table 14). First, the urban sites appeared to have greater 90th centile concentrations than the agricultural sites. Second, the 90th centile concentration for the integrator sites representing larger watersheds was the smallest. Third, the greatest 90th centile concentration for these pooled data did not exceed 25 ng/L.

Comparison of Data Sets from the Same Geographic Areas. Data from the Lake Erie drainage basin study were more extensive than comparable data from

Table 10. Nonlinear regression coefficients for interpolated 2-d time-averaged monitoring data from Lake Erie drainage basin watersheds.

Watershed	Years monitored	Mean no. daily observations/yr	Interpolated N	r^2	a	b	c	d
Lost Creek, OH	1893–1900, 1992–1993	52	3464	0.998	86.54	1.14	8.75	100.31
Rock Creek, OH	1983–1995	64	4493	0.998	80.01	1.00	12.14	101.36
Honey Creek, OH	1983–1995	70	4494	0.999	86.59	1.23	13.30	100.57
Huron River, OH	1988–1991	48	1188	0.996	69.65	0.82	15.23	99.78
River Raisin, MI	1983–1995	64	4501	0.996	83.17	0.83	14.34	100.61
Sandusky River, OH	1983–1995	67	4501	0.996	80.84	0.58	19.42	99.83
Maumee River, OH	1983–1995	64	4501	0.996	83.17	0.83	14.34	100.61

Table 11. Chlorpyrifos 90th centile 2-d time-weighted mean concentrations in Lake Erie drainage basin watersheds.

Watershed	Area (km^2)	90th centile concentration (ng/L)
Lost Creek, OH	11	3
Rock Creek, OH	88	11
Honey Creek, OH	390	5
Huron River, OH	930	37
River Raisin, MI	2,700	8
Sandusky River, OH	3,200	17
Maumee River, OH	16,000	8

NAWQA. Moreover, the Lake Erie drainage basin study utilized event-driven stratified temporal sampling and was a long-term project, whereas NAWQA sampling tended to be restricted to calendar scheduling over a relatively short time frame. Ninetieth centile concentrations from the Lake Erie drainage basin data were compared to those from the nearest corn belt NAWQA study unit, the White River Basin in Indiana, to determine the degree of agreement in monitoring results between these dissimilar sampling programs on similar watersheds in the eastern portion of the Midwestern U.S.. The mean of the 90th centiles of the 2-d time-weighted average concentrations for the Lake Erie watershed locaions (Table 11) was 12.7 ng/L, while the mean of the 90th centiles in the four White River ("whit-") locations (Table 14) was 19.8 ng/L. Assuming that the two watersheds were indeed similar in behavior during the respective monitor-

Table 12. Greatest annual daily maximum measured concentrations of chlorpyrifos in surface waters of the Lake Erie drainage basin watershed study (ng/L).

Year	Maumee	Sandusky	Raisin	Huron	Honey	Rock	Lost	Cuyahoga
1983	0	0	0	—	0	0	0	0
1984	0	0	0	—	0	0	0	0
1985	0	0	0	—	0	0	0	0
1986	0	0	0	—	0	0	0	0
1987	100	—	90	—	178	—	151	75
1988	282	—	251	546	34	396	96	500
1989	60	225	0	942	84	61	120	46
1990	46	2,456	—	457	85	333	161	80
1991	482	1,552	—	380	76	792	—	39
1992	1,111	1,308	64	—	300	35	531	—
1993	221	2,828	26	—	190	224	662	39
1994	221	1,150	0	—	189	171	—	0
1995	176	333	93	—	226	237	—	86

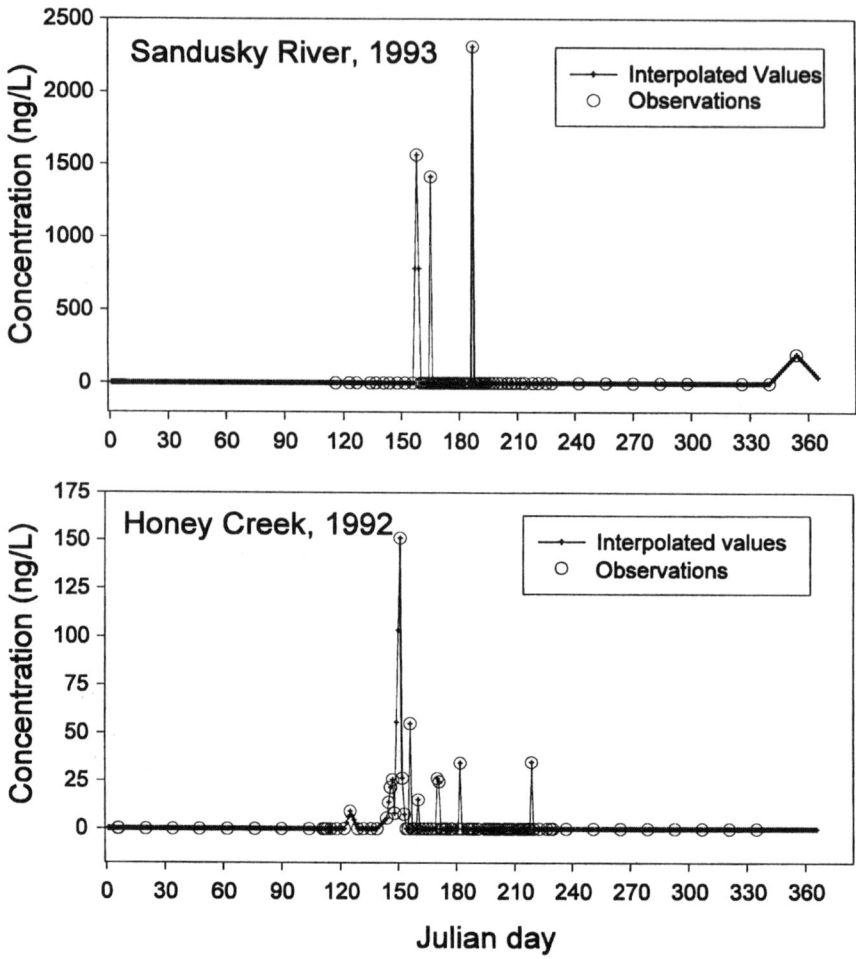

Fig. 14. Daily mean concentrations of chlorpyrifos at two Lake Erie drainage basin locations for the year of reported maximum concentration (Table 12).

ing periods, the NAWQA results appear to be comparable to the Lake Erie drainage basin results. The mean value for the Lake Erie drainage basin data set may be influenced by less chlorpyrifos use in the earlier years of observations.

A comparison was also made between the NAWQA San Joaquin-Tulare Basins study unit and the USGS/DPR California monitoring sites. In this case, generally similar numbers of samples were collected but not necessarily in the same years (Tables 8 and 9). Mean 90th centile concentrations for the locations Merced River, Orestimba Creek, Salt Slough, and San Joaquin River at Vernalis were 112.8 and 48.0 ng/L for the USGS/DPR and NAWQA data sets, respectively (Tables 13 and 14). The difference between the two means is greater than

Table 13. Linear regression coefficients and chlorpyrifos 90th centile concentrations for California data.

Location	N	r^2	a	b	90th centile concentration[a] (ng/L)
Salt Slough	10	1.000	0.396	5.513	88
Orestimba Creek	19	0.958	1.128	3.491	298
Spanish Grant Combined Drain	19	0.918	1.371	2.752	375
TID No. 3	17	0.979	1.011	3.423	672
TID No. 5	22	0.893	1.549	2.951	141
TID No. 6	17	0.882	1.118	3.476	324
Del Puerto Creek	22	0.894	2.109	2.651	53
Ingram/Hospital Creeks	24	0.767	1.076	3.958	145
Merced River	17	0.887	1.275	3.708	104
Tuolumne River	11	0.679	2.708	2.406	27
Stanislaus River	12				10[b]
San Joaquin River at Hills Ferry Road	37	0.774	1.721	4.54	10
San Joaquin River at Patterson	26				ND[c]
San Joaquin River at Laird-Park	122	0.912	1.000	5.094	15
San Joaquin River at Airport Road	14	0.806	2.157	3.099	30
San Joaquin River at Vernalis	1127	0.926	1.694	5.469	≪1
Sacramento River at Sacramento	471				ND[c]
Alamo River at Harris St. Bridge	15	0.879	0.984	3.880	271

[a]Data plotted as concentration on logarithmic scale versus cumulative probability.
[b]Four occurrences of 10 ng/L, eight nondetects.
[c]All concentrations below detection limit. Note that the units for the slope and intercept (a and b) are log$_{10}$ and probits.

the difference between the Lake Erie drainage basin and White River Basin means, but both sampling programs in the San Joaquin Valley found greater concentrations relative to the Midwest. Based on this limited analysis, it was concluded that the NAWQA sampling program demonstrated the ability to characterize chlorpyrifos concentrations in the study unit basins during the period of observation. However, the uncertainty in the concentration distributions was not quantified, and this uncertainty could be quite great in some of the study units, given the limited numbers of samples collected at some locations.

Seasonality of Concentrations. Analysis of the monthly 90th centile concentrations for the Lake Erie drainage calculated from interpolated daily values from all locations (Fig. 15, Table 15) revealed that the greatest concentrations were

Table 14. Linear and nonlinear regression coefficients and chlorpyrifos 90th centile concentrations for NAWQA data.

Location	Type	N	r^2	a	b	c	d	90th centile concentration (ng/L)	Notes
acfb-aycocks	indicator-agriculture	34						<99	99 ng/L = only value > detection limit.
acfb-lime	indicator-agriculture	54						—	Not detected.
acfb-sope	indicator-urban	52	0.952	2.007	−1.458			23	Concentrations plotted on log scale.
albe-albe	indicator-agriculture	34						<11	11 ng/L = only value > detection limit.
albe-devils	indicator-agriculture	27						<10	10 ng/L = only value > detection limit.
albe-pete	indicator-agriculture	35	0.640	3.828	−2.074			8	Concentrations plotted on log scale.
albe-tar	integrator	20						<7	7 ng/L = only value > detection limit.
ccpt-crab.m	indicator-agriculture	21						—	Not detected.
ccpt-crab.rl	indicator-agriculture	33	0.944	1.148	−0.542			39	Concentrations plotted on log scale.
ccpt-el68	indicator-agriculture	31	0.960	1.350	−0.820			36	Concentrations plotted on log scale.
ccpt-palouse	indicator-agriculture	40						—	Not detected.
cnbr-maple	indicator-agriculture	59	0.981	69.51	4.004	21.04	97.65	27	Concentrations plotted on log scale.
cnbr-platte	integrator	59	0.964	0.938	0.072			19	Concentrations plotted on log scale.
cnbr-prairie	indicator-agriculture	16	0.819	0.552	0.247			75	Concentrations plotted on log scale.

Chlorpyrifos in Aquatic Environments 47

site	type						comments
cnbr-shell	indicator-agriculture	15	0.927	0.943	−0.444	68	Concentrations plotted on log scale.
conn-norwalk	indicator-urban	51				—	Not detected.
gafl-lafayette	indicator-urban	41	0.978	0.103	−0.216	15	Concentrations plotted on linear scale.
gafl-little	indicator-agriculture	37				<21	21 ng/L = only value > detection limit.
gafl-tucsa	indicator-agriculture	53	0.905	0.154	0.792	3	Concentrations plotted on linear scale.
gafl-withla	integrator	37	0.917	0.189	0.374	5	Concentrations plotted on linear scale.
hdsn-canoj	indicator-agriculture	21				—	Not detected.
hdsn-lisha	indicator-urban	22				<9	9 ng/L = only value > detection limit.
hdsn-moh	integrator	20				—	Not detected.
lsus-cedar	indicator-urban	61	0.928	0.039	1.065	6	Concentrations plotted on linear scale.
lsus-eastm	indicator-agriculture	44	0.951	1.411	−0.262	12	Concentrations plotted on log scale.
lsus-mill	indicator-agriculture	43	0.943	1.550	−0.194	9	Concentrations plotted on log scale.
nvbr-lasvegas	indicator-urban	40	0.927	0.785	0.093	33	Concentrations plotted on log scale.
ozrk-dousin	indicator-agriculture	22				—	Not detected.
ozrk-yocum	indicator-agriculture	23				—	Not detected.
poto-accotink	indicator-urban	30	0.978	43.31	3.385	18	Concentrations plotted on log scale.

(*Continued*)

Table 14. Continued.

Location	Type	N	r^2	a	b	c	d	90th centile concentration (ng/L)	Notes
poto-mono	indicator-agriculture	36	0.870	0.756	0.650			7	Concentrations plotted on log scale.
poto-muddy	indicator-agriculture	34						—	Not detected.
poto-shenan	integrator	18						—	Not detected.
redn-rr.em	integrator	22						—	Not detected.
redn-rr.fargo	integrator	10						—	Not detected.
redn-snake	indicator-agriculture	27	0.966	0.022				9	Concentrations plotted on linear scale.
redn-turtle	indicator-agriculture	30						—	Not detected.
redn-wildr	indicator-agriculture	27						—	Not detected.
sanj-merced	indicator-agriculture	50	0.973	35.66	3.584	28.09	95.80	52	Concentrations plotted on log scale.
sanj-orest	indicator-agriculture	85	0.983	1.639	-1.863			83	Concentrations plotted on log scale.
sanj-salt	indicator-agriculture	26	0.951	0.045	-0.297			35	Concentrations plotted on linear scale.
sanj-sj	integrator	73	0.954	0.089	-0.697			22	Concentrations plotted on linear scale.
splt-cherry	indicator-urban	36	0.971	0.898	-0.496			95	Concentrations plotted on log scale.
splt-lone	indicator-agriculture	35	0.988	74.21	4.581	14.47	96.07	18	Concentrations plotted on log scale.
trin-chamb	indicator-agriculture	28						—	Not detected.
trin-rush	indicator-urban	22	0.980	2.008	-2.517			78	Concentrations plotted on log scale.

Site	Type	n					Notes
usnk-rock	indicator-agriculture	41	1.000	1.408	0.573	3	Concentrations plotted on log scale.
usnk-teton	indicator-agriculture	29				—	Not detected.
whit-kess	indicator-agriculture	45	0.776	0.013	0.845	35	Concentrations plotted on linear scale.
whit-little	indicator-urban	96	0.952	2.225	−1.907	27	Concentrations plotted on log scale.
whit-sugar	indicator-agriculture	72	0.973	1.182	0.306	7	Concentrations plotted on log scale.
whit-white	integrator	72	0.864	1.173	0.089	10	Concentrations plotted on log scale.
will-fanno	indicator-urban	28	0.971	0.052	−0.491	34	Concentrations plotted on linear scale.
will-pudding	indicator-agriculture	28	0.981	1.561	−0.473	13	Concentrations plotted on log scale.
will-zollner	indicator-agriculture	29	0.987	34.39	2.469	37	Concentrations plotted on log scale. 16.49 97.38
wmic-duck	indicator-agriculture	42				<10	10 ppt = only value > detection limit.
wmic-milw	integrator	34				<16	16 ppt = only value > detection limit.
wmic-nbmilw	indicator-agriculture	36				—	Not detected.
All sites	—	2,194	0.977	1.244	−0.214	16	Concentrations plotted on log scale.
All indicator-agriculture	indicator-agriculture	1342	0.970	1.108	0.000	14	Concentrations plotted on log scale.
All indicator-urban	indicator-urban	479	0.970	1.497	−0.787	24	Concentrations plotted on log scale.
All integrator	integrator	373	0.956	1.437	−0.227	11	Concentrations plotted on log scale.

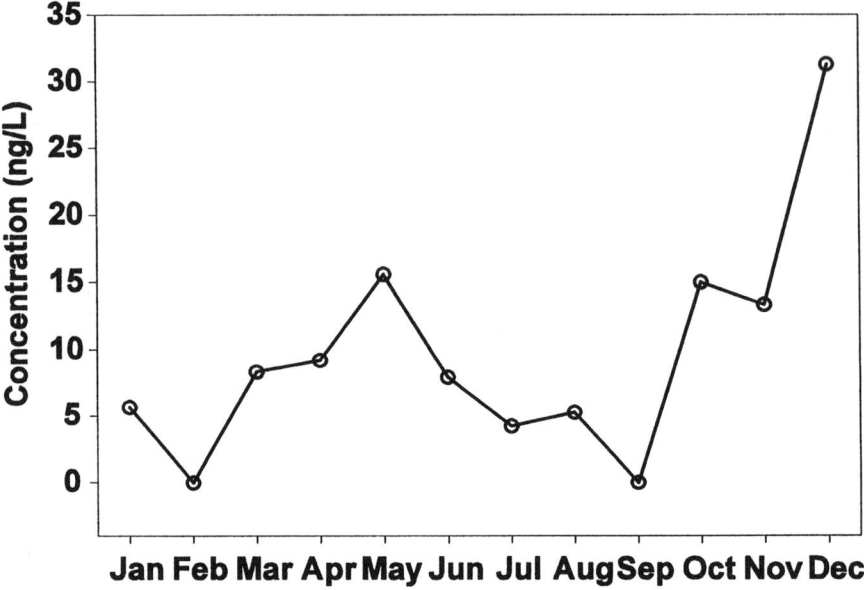

Fig. 15. Ninetieth centile monthly chlorpyrifos concentrations estimated from pooled daily interpolated data from locations in the Lake Erie drainage basin.

Table 15. Nonlinear regression coefficients and chlorpyrifos 90th centile concentrations for monthly Lake Erie drainage basin interpolated data.

Month	N	r^2	a	b	c	d	90th Centile concentration[a] (ng/L)
Jan	1860	0.994	84.28	1.11	9.46	100.1	6
Feb	1695	0.984	92.56	1.77	5.20	99.7	0[b]
Mar	1860	0.989	83.71	0.84	18.6	102.3	8
Apr	1920	0.998	85.35	0.90	28.3	102.7	9
May	2022	0.999	78.74	1.27	15.0	100.8	16
Jun	1980	0.996	84.73	1.10	14.3	100.1	8
Jul	2046	0.999	88.54	1.00	29.4	100.2	4
Aug	1917	0.984	86.19	0.79	15.7	99	5
Sep	1827	0.969	92.01	1.00	5.52	98.9	0[b]
Oct	1860	0.990	77.50	0.66	10.0	99.5	15
Nov	1800	0.996	74.95	0.83	7.72	99.5	13
Dec	1860	0.997	73.88	0.81	20.0	101.2	31

[a] Concentrations plotted on log scale.
[b] 90th centile plotted value less than detection limit.

observed in December, with a secondary peak in May. These data are difficult to interpret, but the greater concentrations in the early winter (and fall) are unlikely to be related to the use of granular chlorpyrifos in preplant applications to field corn. The May peak is consistent with runoff events associated with at-plant corn applications.

In the San Joaquin River drainage, monthly 90th centile concentrations were estimated for the USGS Vernalis data set and also for the pooled data from all locations upstream of Vernalis (Fig. 16, Table 16). At the "integrator" site of Vernalis, concentrations were less than the detection limit in all months except for February and March, when small concentrations of chlorpyrifos were measured. Greater concentrations were observed in the pooled tributary data from January through April, with the peak of 178 ng/L reported in March. In the second half of the year concentrations were at or less than the method detection limit. This pattern is consistent with use of chlorpyrifos as a dormant spray in orchards and early-season use in alfalfa, when natural rainfall can result in large-scale runoff events and large flow volumes in the river system. Flows from April through October are much lower and resulted primarily from agricultural drainage under irrigated conditions. The small concentrations observed at Vernalis relative to those found in tributaries are attributable to dilution, dispersion, and relatively rapid dissipation of chlorpyrifos from the water column.

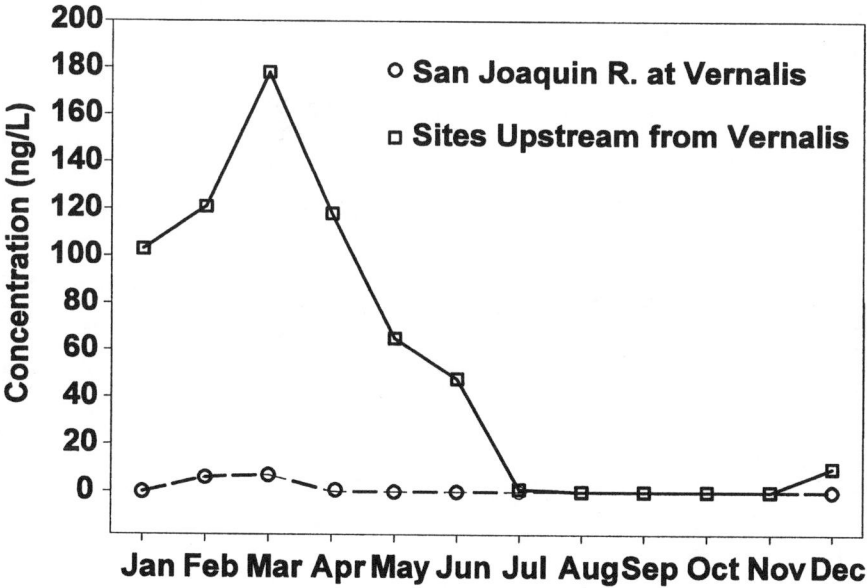

Fig. 16. Ninetieth centile monthly chlorpyrifos concentrations for the San Joaquin River at Vernalis and for pooled data from upstream locations.

Table 16. Linear regression coefficients and chlorpyrifos 90th centile concentrations for monthly San Joaquin River data.

Location	Month	N	r^2	a	b	90th Centile concentration (ng/L)	Notes
San Joaquin River at Vernalis	Jan	110				0	All concentrations below detection limit.
	Feb	109	0.962	1.291	0.247	6	Concentrations plotted on log scale.
	Mar	115	0.639	2.047	−0.502	7	Concentrations plotted on log scale.
	Apr	114				0	All concentrations below detection limit.
	May	84				0	Two occurrences of 12 ng/L.
	Jun	80				0	Two occurrences of 22 ng/L.
	Jul	86				0	Two occurrences of 15 ng/L.
	Aug	90				0	All concentrations below detection limit.
	Sep	84				0	All concentrations below detection limit.
	Oct	81				0	All concentrations below detection limit.
	Nov	86				0	All concentrations below detection limit.
	Dec	87				0	All concentrations below detection limit.
Sites upstream from Vernalis[a]	Jan	39	0.766	0.816	−0.362	103	Concentrations plotted on log scale.
	Feb	43	0.946	1.266	−1.354	121	Concentrations plotted on log scale.
	Mar	46	0.956	1.064	−1.114	178	Concentrations plotted on log scale.
	Apr	51	0.934	1.169	−1.139	118	Concentrations plotted on log scale.
	May	63	0.869	1.571	−1.569	65	Concentrations plotted on log scale.
	Jun	45	0.733	1.094	−0.557	48	Concentrations plotted on log scale.
	Jul	26	>0.999	0.183	1.264	1	Concentrations plotted on log scale.
	Aug	36				0	One occurrence of 350 ng/L.
	Sep	22				0	One occurrence of 330 ng/L.
	Oct	5				0	All concentrations below detection limit.
	Nov	2				0	Both concentrations below detection limit.
	Dec	11				10	Two occurrences of 10 ng/L.

[a] Orestimba Creek, Spanish Grant Combined Drain, TID no 3, 5 and 6, Del Puerto Creek, Ingram/Hospital Creeks, Merced River, Tuolomne River, Stanislaus River, and San Joaquin River at Hills Ferry Road, at Patterson, at Laird Park, and at Airport Road.

Pulse Width. Chlorpyrifos typically appears in monitoring data as a series of distinct pulses characterized by individual peak concentrations that may or may not be flanked by lesser concentrations producing a pulse of varying width. A temporal description of the exposure profile is particularly important for a toxicant such as chlorpyrifos that is acutely toxic with a small acute-to-chronic ratio (ACR). Therefore, all the available data sets were evaluated for their potential to characterize the duration of exposure. Only the Lake Erie drainage and the USGS Sacramento River and San Joaquin River at Vernalis monitoring studies reported sufficient numbers of observations for this analysis. The USGS data, however, contained too few detections to characterize typical pulse width. The stratified sampling scheme applied to the Lake Erie drainage basin monitoring program provided adequate resolution in time during the annual runoff period of interest, and these data were selected for further analysis.

Chlorpyrifos concentrations in the samples from the Lake Erie drainage basin data sets (raw data, not daily averages or interpolations) were reviewed for instances in which sufficient numbers of samples were collected to characterize consecutive-day appearances of chlorpyrifos in surface water. It was hypotheized that the typical pulse width for these data was 48 hr or shorter, and the data were therefore examined for cases where the observed pulse width was greater than this duration (Table 17). The overall frequency of observing a pulse greater than 48 hr was about 0.135; thus, more than 85% of the reported chlorpyrifos pulses lasted 2 d or less. Adoption of 48 hr as the typical pulse width is therefore reasonable. The bias introduced could be small, and the potential to underestimate the duration of exposure would be only 15%.

Geographic Distribution of 90th Centile Concentrations. Ninetieth centile concentrations for all sites in the U.S. (Fig. 17) were interpreted in relation to the geographic use data presented in Figs. 5–8 and vulnerability to runoff losses. Agricultural uses (Fig. 5) are distributed over certain distinct regions that are clearly determined by the combination of crop use sales volume (see Table 3) and county crop statistics reported in the 1992 Census of Agriculture (Bureau of the Census, U.S. Department of Commerce). These regions are given in Table 18, which also summarizes the relative magnitudes of the observed 90th concentrations in the regions. Sites with the greatest concentrations were located in California, where use in dormant trees predominates during the rainy winter.

Table 17. Numbers of pulse widths greater than 40 hr observed in the Lake Erie drainage basin consecutive day sampling.

	Lost Creek	Rock Creek	Honey Creek	Huron River	River Raisin	Sandusky River	Maumee River	All
No. > 48 hr	1	3	2	2	—	3	1	12
N	7	17	23	4	0	25	13	89

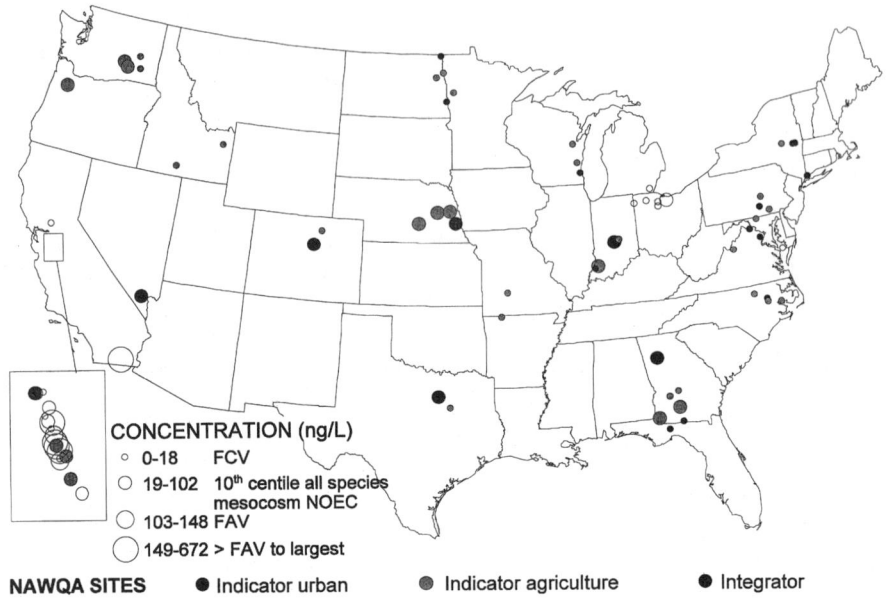

Fig. 17. Graduated symbol map of 90th centile concentrations of chlorpyrifos at all sampling sites.

Table 18. Major crop regions associated with high-volume use of chlorpyrifos and relative magnitude of 90th centile concentrations (agricultural and integrator sites).

Region	Study areas	Primary crop	Relative concentration
Midwest	WHIT, CNBR, WMIC, Lake Erie	Field corn	Small to moderate
Red River Valley	REDN	Sugar beets	Small
High Plains	SPLT	Wheat	Small
Mid-Atlantic	ALBE	Tobacco	Small
Southest	ACFB, GAFL	Peanuts	Small to moderate
California	SANJ, California sites	Dormant trees, foliar nuts, alfalfa, mixed row crops	Moderate to large

WHIT, White River basin, IN.; CNBR, Central Nebraska basin; WMIC, Western Lake Michigan basin; REDN, Red River of the North basin; SPLT, South Platte basin; ALBE, Albemarale-Pamlico basin; ACFB, Apalachicola-Chattahoochee-Flint River basin; GAFL, Georgia-Florida coastal basin; SANJ, San Joaquin-Tulare basins

During the irrigation season, use is spread over several tree and field crops. Smaller concentrations were found in the southeastern peanut region and in the corn belt, while Red River sugar beets and Mid-Atlantic tobacco were associated with the fewest chlorpyrifos detections. Data from the one site in the High Plains wheat area also suggests that small concentrations may be typical of this use pattern as well. No information was available for cotton use in the Mississippi Delta nor for citrus in California, Florida, or Texas.

Specialty uses (see Fig. 6) were associated with urban counties, states with larger populations, and the eastern seaboard. Chlorpyrifos use coincided with sampling at all 10 NAWQA urban indicator sites, but there was no obvious relationship between use volume and the magnitude of reported concentrations. This was not unexpected, given the small number of urban indicator sites and the uncertainties in the distribution of specialty products described earlier. The three sites in the northeast reported the smallest concentrations. The combined sales map (see Fig. 7) provides a more complete portrayal of the density of total chlorpyrifos use on an area basis. However, it was not possible to glean any additional interpretations of the monitoring data from examining this view of product distribution.

There was no clear relationship between the observed chlorpyrifos concentrations and potential runoff vulnerability based on a 30-yr average of weather patterns (Fig. 8). Chlorpyrifos use appeared to be a more reliable predictor of the magnitude of concentration. This was observed in the less vulnerable Nebraska study unit (CNBR), where chlorpyrifos use was greater than in the more vulnerable Indiana (WHIT) and Lake Erie watersheds (Fig. 5). Another example is the peanut-growing region in southern Georgia. Some of the chlorpyrifos detections occurred in areas with zero or low vulnerability to potential runoff under natural rainfall or in areas that were not simulated. It is probable that the chlorpyrifos residues detected at these sites were transported by irrigation tailwater or by other processes related to chlorpyrifos use (e.g., spray drift, improper disposal). Again, the greater concentrations tended to be associated with areas of greater agricultural use of chlorpyrifos (SANJ, Alamo River).

Analysis of the Saltwater Data. Greatest concentrations of chlorpyrifos in Chesapeake Bay were observed in March and April with lesser concentrations being observed in June and September. The greatest concentration observed in the year was 1.67 ng/L (McConnell et al. 1997). Although the number of data points in the study was small and they represented a spatial rather than a time series, the data were analyzed in an attempt to assess risks from chlorpyrifos in estuarine environments. Because the data sets were small, the log-normal distribution was assumed. Regressions from the data in Fig. 18 were used to calculate the 90th centiles (Table 19). Compared to other sampling sites in freshwater systems, the 90th centiles were low, possibly because of further dilution of stream and river runoff residues in the estuarine environment.

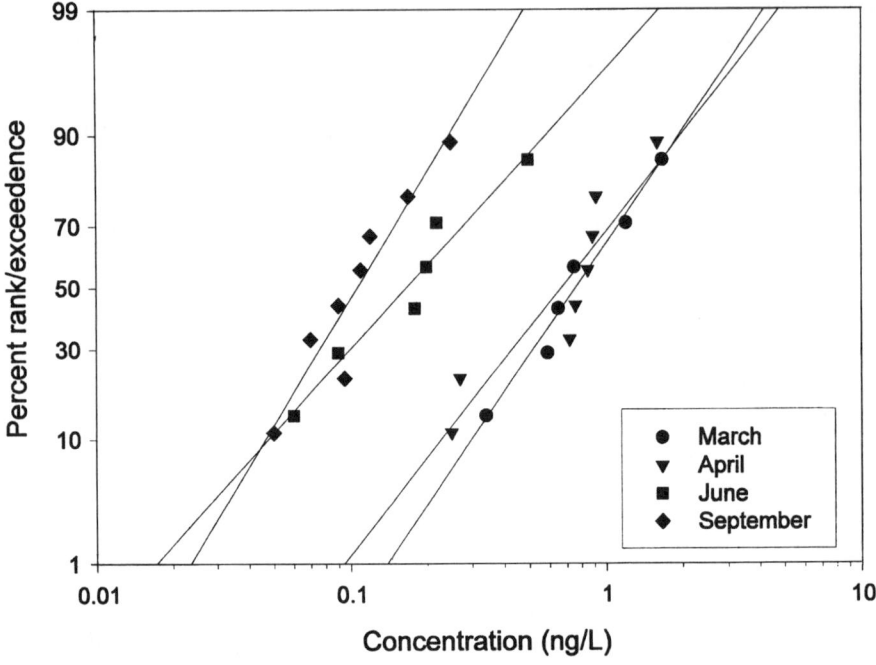

Fig. 18. Distributions of chlorpyrifos concentration data from Chesapeake Bay.

IV. Effects Characterization
A. Bioconcentration and Bioaccumulation

Bioconcentration Potential. Fish and other aquatic organisms rapidly absorb chlorpyrifos from water, with reported uptake rates ranging from 2 to 60 mL/g^{-1}/hr^{-1} and elimination half-lives of 0.5–4 d (Barron and Woodburn 1995; Montañés et al. 1995). The bioconcentration of chlorpyrifos is usually less than pre-

Table 19. Regression coefficients and chlorpyrifos 90th centile concentrations for the Chesapeake Bay data.

Month	N	r^2	a	b	90th centile (ng/L)
March	6	0.97	3.15	5.37	1.9
April	8	0.85	2.73	5.47	2.0
June	6	0.95	2.35	6.83	0.6
September	8	0.89	3.54	8.44	0.2

Note that the units for the slope and intercept (a and b) are $\log_{10}$ and probits.

dicted from equilibrium partitioning models, chiefly due to biotransformation of both waterborne and dietary chlorpyrifos. Measured bioconcentration factors are in the range of 50 to 6000 (Barron et al. 1991; Deneer 1994; Neely et al. 1974; Welling and de Vries 1992). However, extensive partitioning onto organic matrices may also limit the bioconcentration of chlorpyrifos. For example, the presence of organic materials (i.e., plants, sediment) in the exposure water limited bioaccumulation in goldfish, due to adsorption of chlorpyrifos (Smith et al. 1966). The bioconcentration of the degradation products of chlorpyrifos, such as TCP, are expected to be less than the parent because of greater polarity. For example, the measured bioconcentration factor of TCP in surviving mosquitofish was approximately 100 times less than that of chlorpyrifos (Neely et al. 1974).

Bioconcentration Factors. Bioconcentration factors (BCFs) for chlorpyrifos accumulation by invertebrates and fish range from 42 to 5100 mL/g wet wt, depending on the species, exposure concentration, and exposure conditions (Barron and Woodburn 1995; Montañés et al. 1995); an average BCF value of approximately 1000 mL/g is generally used in aquatic risk assessment calculations for chlorpyrifos. Any concentration dependence of chlorpyrifos bioconcentration may be caused by toxicity at greater exposure concentrations. For example, Deneer (1993, 1994) suggested that elevated bioconcentration in guppies and sticklebacks at greater than sublethal concentrations may have been caused by inhibition of chlorpyrifos biotransformation, which, in turn, lead to slower elimination. Similarly, Montañés et al. (1995) noted a significant reduction in both uptake and elimination rates of chlorpyrifos with a freshwater isopod (*Asellus aquaticus*) at ambient concentrations of 5 µg/L compared to 0.7 µg/L. The authors measured average uptake and elimination rates of approximately 30 mL/g^{-1}/hr^{-1} and 0.02 /hr, respectively. These estimates of kinetic values would result in a steady-state BCF value of approximately 1500 mL/g.

Biotransformation. The capacity and pathways for chlorpyrifos biotransformation are species specific. For example, Metcalf (1974) reported that algae, snails, and mosquitos exposed to chlorpyrifos in a laboratory ecosystem contained the parent compound, TCP, and unidentified polar metabolites. Oysters biotransformed chlorpyrifos to *O,O*-diethyl-*O*-(3,5-dichloro-6-methylthio-2-pyridyl) phosphorothioate (Hansen et al. 1992). This pathway appears to occur in bivalves, but generally not in fish. Fish biotransform chlorpyrifos to a variety of metabolites, including TCP, methoxytrichloropyridine, and glucuronide conjugates (Barron et al. 1991). For example, goldfish produced at least four metabolites of chlorpyrifos. TCP was the principal compound in the exposure water of goldfish 5 d after an initial exposure to 50 µg/L chlorpyrifos (Smith et al. 1966). Mosquitofish exposed to chlorpyrifos in a laboratory ecosystem contained 50% parent, 29% TCP, and 21% unidentified polar metabolites (Metcalf 1974). Mosquitofish have been reported to metabolize chlorpyrifos to TCP and two unidentified polar metabolites (Hedlund 1973). Rainbow trout biotransformed chlorpyrifos to TCP and unidentified conjugates (Murphy and Lutenski 1986). Excretion of parent chlorpyrifos by guppies was negligible, which indicates that biotransformation was the dominant elimination pathway (Welling and de Vries 1992). In channel catfish, excretion of parent chlorpyrifos was also limited, and biotransformation was the dominant elimination pathway (Barron et al. 1993).

B. Toxicity of Metabolites

Some data are available to address the potential toxicological significance of transformation products. The literature indicates that none of the transformation products of chlorpyrifos are sufficiently persistent, bioaccumulative, or toxic to be of toxicological concern. Our analysis of the relative concentrations of metabolites and chlorpyrifos toxicities indicated that protection of organisms from the effects of chlorpyrifos should protect them from any adverse effects of the metabolites. TCP does not cause cholinesterase inhibition and is several orders of magnitude less toxic to aquatic organisms than chlorpyrifos. Typical laboratory 96-hr LC_{50} values for TCP include 12,500,000 ng/L for bluegill sunfish (*Lepomis macrochirus*), 12,600,000 ng/L for rainbow trout (*Oncorhyncus mykiss*), and 10,400,000 ng/L for daphnids (*Daphnia magna*) (Gorzinski et al. 1991a–c). The 96-hr LC_{50} value for TCP of 83,000,000 ng/L with the grass shrimp (*Palaemonetes pugio*) and a 96-hr EC_{50} (inhibition of shell growth) with the eastern oyster (*Crassostrea virginica*) of 9,300,000 ng/L (Graves and Smith 1991a,b). The 96-hr LC_{50} in six species of salmonids ranged from 1,800,000 to 2,700,000 ng/L for TCP and from 1,100,000 to 6,300,000 ng/L for the minor chlorpyrifos degradate methoxytrichloropyridine (Wan et al. 1987). In the Microtox^{TM} microbial bioassay, the trichloropyridinol had an LC_{50} of 18,600,000 ng/L to a marine *Photobacterium* bacterium, indicating an effect only at a high, environmentally unrealistic concentration (Somasundaram et al. 1991). There are no indications that TCP causes any acute, chronic, or secondary toxicity in the environment (Racke 1993).

Chlorpyrifos oxon, the activated form that is the most effective inhibitor of acetylcholinesterase *in vivo*, is acutely toxic to animals such as rodents and fish, and is therefore of potential importance to nontarget species if it ever occurred in the environment in significant quantities. The oxon is extremely sensitive to hydrolytic degradation and only occurs at small concentrations in environmental media and is not detectable in tissues of any organisms collected in the environment (Racke 1993). LC_{50}s were determined for both chlorpyrifos and the oxon in four age groups of the medaka. The 48-hr LC_{50}s of the oxon were, on average, 13 fold less than for chlorpyrifos. The range of LC_{50}s was 14,000–33,000 ng/L for the oxon and 184,000–404,000 µg/L for chlorpyrifos (Rice 1992). Body burdens for chlorpyrifos after 48-hr exposure have been reported as 31–463 µg/g (Rice et al., 1997). After medaka were exposed to chlorpyrifos oxon for 48 hr, no residues of the oxon were detectable in the tissues, a reflection of its very short biological half-life (Rice et al., 1997). In light of the very low concentrations and no measurable persistence in environmental matrices, the oxon is not considered to be of any toxicological importance in an ecological risk assessment.

There is no information on adverse effects from the dealkylated products. The other metabolites that occur in the environment (dealkylation products, methoxytrichloropyridine, ethyl phosphates, and thiophosphates) are also fully biodegradable and nontoxic at environmentally relevant concentrations.

C. Pulsed Exposures

Although pulsed exposures commonly occur in aquatic ecosystems, limited information is available for evaluating the effects of pulsed exposures relative to continuous exposures. Profiles for concentration versus pulse time in toxicity testing have been addressed for chlorpyrifos in laboratory and mesocosm studies, using several aquatic species. Laboratory studies with the fathead minnow have shown short-duration pulses of exposure to have less impact than continuous exposure with the same product of duration of exposure multiplied by exposure concentration (area under the curve), unless the concentrations are very much greater (Jarvinen et al. 1988). Outdoor stream mesocosm experiments with bluegill and fathead minnow demonstrated that high-concentration pulses caused a greater peak cholinesterase inhibition than a continuous low-concentration exposure. However, pulsed exposures also resulted in more reversible effects than continuous exposures (Eaton et al. 1985). Observations on mosquitofish accidentally poisoned with chlorpyrifos showed recovery of brain AChE activity from >70% inhibited to activity similar to control fish in 45 d (Carr et al. 1997). Pulsed exposures have also been evaluated in *Daphnia magna*. For 6- or 12-hr pulses (at 1.0 or 0.5 mg/L, respectively), the daphnids recovered so long as there was a 72-hr interval between pulses. Minimal effects were noted in general for short-term pulses, compared to fewer, but long-term, pulses (Naddy 1996).

D. Aquatic Ecotoxicology

The ecotoxicology of chlorpyrifos in terrestrial and aquatic systems has recently been summarized in a literature review by Barron and Woodburn (1995). In brief, the sensitivity of aquatic species to chlorpyrifos varies considerably among kingdoms, phyla, and species. Laboratory tests have shown that most species of aquatic plants (including algae) are relatively tolerant of chlorpyrifos exposure, with EC_{50}s (3–5 d) generally >100,000 ng/L. Microcosm and field studies have demonstrated that algal populations are affected by chlorpyrifos exposure primarily through secondary effects from changes in the abundance of grazer populations (e.g., zooplankton). Aquatic invertebrates exhibit species-specific variation in their tolerance of exposure to chlorpyrifos. For example, LC_{50}s for chlorpyrifos in aquatic invertebrates range from 1.0 ng/L (mosquito larvae) to >12,000,000 ng/L (rotifer species). In general, crustaceans and insect larvae are the most sensitive aquatic species, with LC_{50}s <10,000 ng/L for some species. Rotifers, mollusks, and annelids are generally the least sensitive aquatic invertebrates in both laboratory and field studies. Chronic laboratory life cycle exposures (21–35 d) of invertebrates to constant concentrations of chlorpyrifos have shown effects on reproduction at concentrations of 4 ng/L (mysid shrimp) to 100 ng/L (daphnids) in laboratory studies. Field studies have shown rapid recovery of invertebrate populations when chlorpyrifos concentrations decline.

In general, fish are less sensitive to chlorpyrifos in both acute and chronic exposures than are invertebrates. Reported laboratory $LC_{50}s$ (96-hr) for two of the most sensitive fish species ranged from 400 ng/L for the saltwater silverside to 2000 ng/L for the freshwater bluegill sunfish. Decreased fish survival at >1000 ng/L and effects on fish growth at 500 ng/L have been observed in field studies with chlorpyrifos. In general, saltwater and freshwater organisms exhibit similar sensitivities to chlorpyrifos.

Reciprocity. The assessment of risk involves determining the probability of the concentration to which organisms are exposed exceeding the concentration of toxicant that will cause an unwanted effect, as determined from the dose–response relationship (Solomon et al. 1996). The degree of response is a function of duration and intensity of exposure. At greater concentrations of toxicant, shorter times of exposure are required to generate a specific response, and vice versa, and this is commonly a reciprocal relationship. The reciprocity between concentration (magnitude) of toxicant and time (duration) to an effect can be used to describe toxicity curves (Giesy and Graney 1989) (Fig. 19).

In the environment, organisms are exposed to fluctuating concentrations that are episodic in nature, whereas in toxicity tests organisms are usually exposed to constant concentrations of toxicant for specified periods of time, such as 48 or 96 hr. The most appropriate measure of exposure to use in a risk assessment is a function of the type of toxicant exposure expected in the field as well as data availability. It is necessary to determine the most relevant type of exposure, acute or chronic, and then scale the measurement of toxicity to these exposures. To do this, one must determine the concentration—body dose-rate relationships for the compound of interest. These relationships are referred to as the reciprocity function. Basically, the scaling determines the most appropriate scenarios of exposure duration, provided that these are less than the maximum duration of the longest period of observation. For corn uses, ecologically relevant exposure scenarios of chlorpyrifos are short-term, pulsed exposures, which are a function of transport events, primarily as runoff associated with rain storms. This, coupled with the fact that the acute-to-chronic ratio (ACR) for chlorpyrifos is approximately 8 (see Table 22, later in this review), indicates that short-term acute exposures are the most likely to cause adverse effects in aquatic organisms. As most episodic aquatic exposures seem to be of 48-hr or shorter duration, the 48-hr LC_{50} was deemed to be the most appropriate duration of exposure for use in risk characterization (see discussion of pulse width on page 51 & 52).

There are 48-hr LC_{50} values published for more than 100 aquatic species. If the 48-hr LC_{50} values can be estimated for all other species for which acute toxicity data are available, then more appropriate comparisons and rankings of species susceptibility to chlorpyrifos can be achieved. Because reciprocity re-

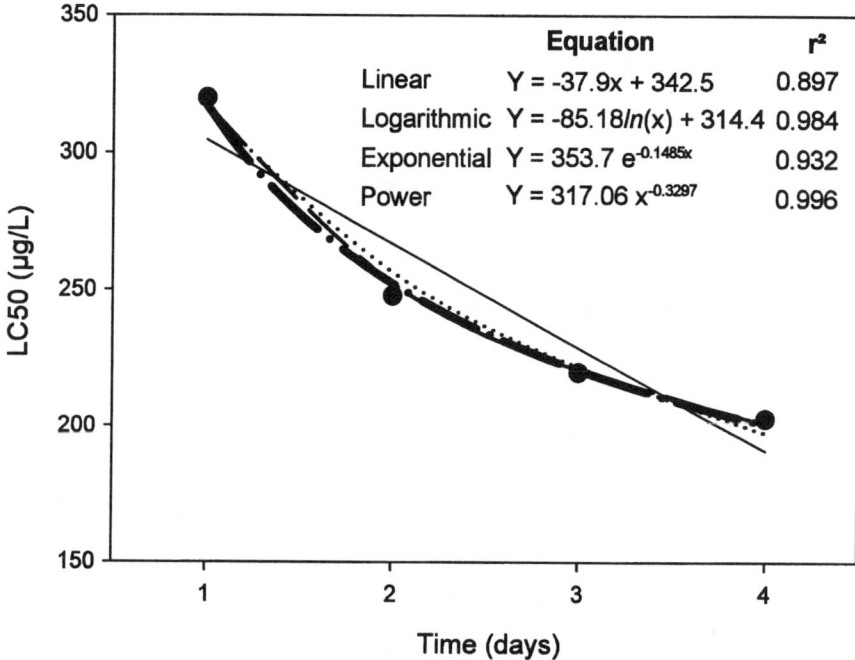

Fig. 19. Example of the relationship between time of exposure and toxicity of chlorpyrifos in fathead minnows.

lates the duration and intensity of exposure to responses, this relationship can be employed to calculate a toxic response at any specified time, and 48-hr LC_{50}s can be estimated from published data for other time periods (e.g., 24 or 96 hr). Strict reciprocity is demonstrated by the situation in which the product of duration and intensity of exposure is a constant. Theoretically, the graphical relationship between duration and intensity of dose is a hyperbola where incipient lethal times and concentrations are approached asymptotically. However, over short ranges of time or concentration, the relationship can often be described by a linear relationship. If strict reciprocity exists, then extrapolation from the available data is straightforward. For instance, if the reciprocity is a direct function, multiplying a 24-hr LC_{50} value by 0.5 would generate an estimated 48-hr LC_{50} for use in the assessment; likewise, a 96-hr LC_{50} could be used to determine an estimated 48-hr LC_{50} by multiplying that 96-hr value by 2. However, if strict reciprocity is observed but the function describing the relationship is not linear, the calculation of the 48-hr LC_{50} values requires closer examination of the relationship between toxicity and exposure time. The results of the detailed analysis of reciprocity indicated that dose–response data from other durations of exposure could be converted to an equivalent 48-hr LC_{50} through the use of a power function ($y = ax^b$, where x is time, y is concentration, and a and b are empirically fitted constants). Therefore, data were converted to 48-hr equivalents and compared to 48-hr moving average concentrations of chlorpyrifos in water.

Toxicity Data. The available LC_{50} data for fish, crustaceans, and insects were used to test the reciprocity relationship between time and toxicity for chlorpyrifos; specifically, LC_{50} data that were available for more than one time interval for a given species were utilized. From those data, examples were only selected for which the same formulation, exposure methodology, and size/age of organisms were used in the toxicity testing; data were therefore only compared for which the LC_{50} values were determined under uniform conditions.

For fish, 17 studies were identified with reported LC_{50} values at more than one time interval, using uniform conditions. These data were plotted for each case and linear regression used to fit a line to the data points. Four studies of chlorpyrifos toxicity to fish reported LC_{50}s for three or more time periods. For these 4 studies, data were plotted using a linear function, exponential function, logarithmic function, and power function. For crustaceans, 14 studies were found that presented LC_{50} values for more than one time period and under uniform conditions. Five of the studies of the effects of chlorpyrifos to crustaceans presented LC_{50} values at more than two time points. For these 5 studies, the LC_{50} values were plotted against time, again using the linear, exponential, logarithmic, and power functions. Of 13 studies on aquatic insects with uniform conditions that determined LC_{50}s, 5 yielded three or more values. These were then selected for further plotting, again using linear, exponential, logarithmic, and power functions as deemed appropriate by least squares analysis based on maximum r^2 (see Fig. 19).

Characterization of the Toxicity Data. The toxicity data were obtained from an extensive review of the literature (Barron and Woodburn 1995) and several other sources (USEPA Pesticide Toxicity Data Base, Montague 1996; AQUIRE,1994). Acute toxicity data were selected to include measures of effects such as LC_{50}s, and EC_{50}s in which the measurement endpoint was clearly related to survival, growth, or reproduction. Biomarker endpoints such as inhibition of cholinesterase, that would normally serve only as indicators of exposure and which are not easily related to assessment endpoints such as survival or reproduction, were not included. Where multiple reports of toxicity were found in the literature, those related to the most sensitive life stage were selected above those for less sensitive stages of the same organism. This added a measure of conservatism to the characterization of the effects. If multiple reports of the same life stage existed, the geometric mean of the observations was calculated. Thus, each species was represented only once in the data set. The characteristics of chlorpyrifos use, environmental fate, and mechanism of action are such that it was judged appropriate to compare acute exposures (0–48 hr) to acute toxicity information from exposures of similar duration. Chlorpyrifos has a relatively small acute-to-chronic ratio of 8 (see Table 22). For this reason, if exposure to chlorpyrifos does not result in acute lethality, it is likely that organisms exposed to short-term pulses will recover.

Toxicity data were grouped for the purposes of assessment and then ranked from least to greatest. Some organisms, such as algae, Mollusca, and rotifers, are tolerant of chlorpyrifos. Data were too few to be analyzed independently and were omitted from most of the analyses. Plotting positions were expressed as percentages (Eq. 1) and linear regressions performed (SigmaPlot 1997). The 10th centile intercepts of the distributions were determined for each taxonomic group (Figs. 20–25 and Table 20). The distribution of 48-hr LC_{50}/EC_{50}s for all freshwater organisms tested (excluding rotifers, mollusks, and other insensitive organisms) demonstrated a wide range of sensitivity spanning five orders of magnitude and gave a 10th centile intercept of 102 ng/L (Table 20, Fig. 20). Freshwater and saltwater vertebrates exhibited a narrower range of sensitivity (Figs. 22 and 25) with 10th centile intercepts of 5358 and 832 ng/L, respectively. It is not known why freshwater and saltwater vertebrates should differ so much in sensitivity. Saltwater organisms are generally more susceptible to organophosphorus pesticides (Hall and Anderson 1993). However, for chlorpyrifos, this observation also may be a function of the small size of the data set for saltwater organisms and the fact that two saltwater fish were very tolerant, resulting in a smaller slope of the regression line (Fig. 25). Arthropods from both freshwater and saltwater were consistently more susceptible than the vertebrates,

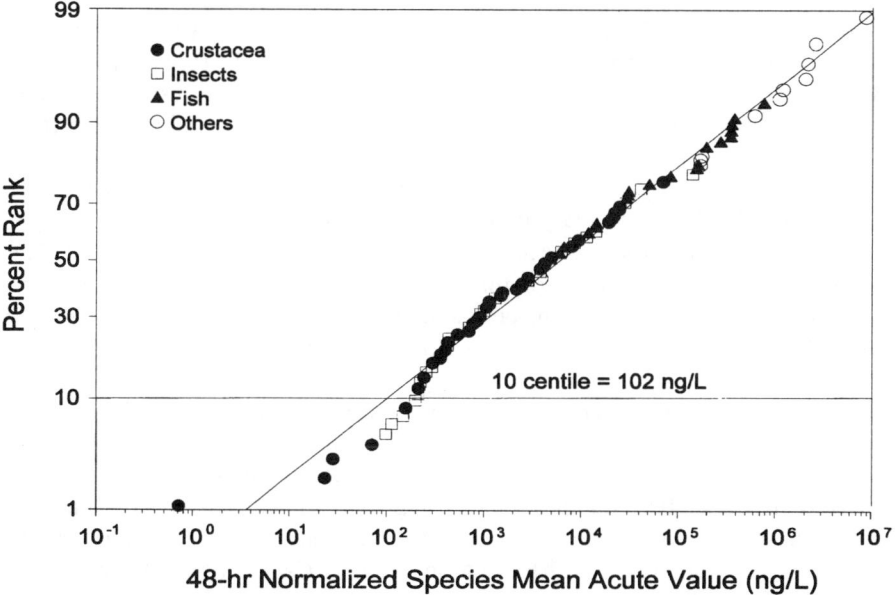

Fig. 20. Distribution of 48-hr normalized EC_{50}/LC_{50} for chlorpyrifos for all freshwater organisms tested.

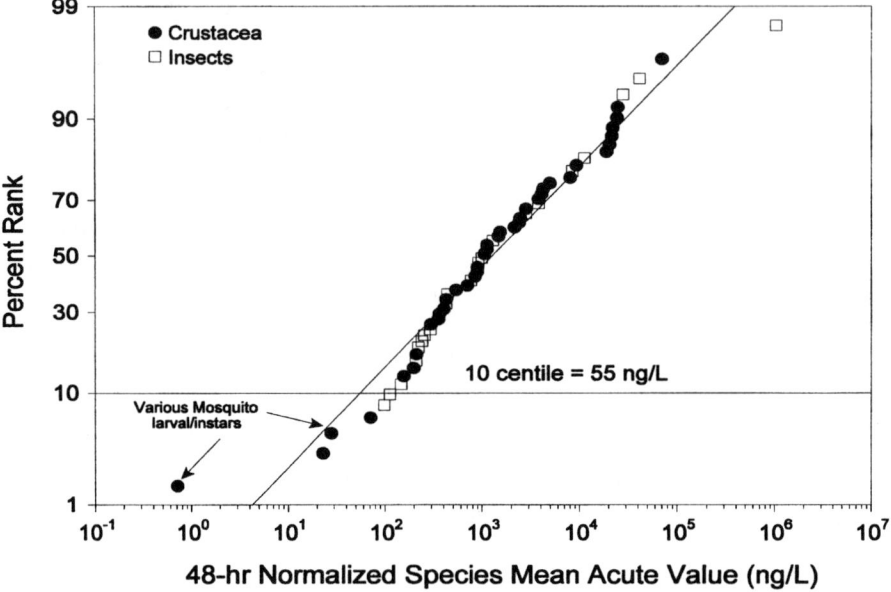

Fig. 21. Distribution of 48-hr normalized EC_{50}/LC_{50} for chlorpyrifos for freshwater arthropods.

and there was a small difference between those from freshwater and saltwater with 10th centile intercepts of 55 and 15 ng/L, respectively (Figs. 21 and 24). Once again, the saltwater data set was smaller than that for freshwater. Some organisms, such as mosquito larvae, were very sensitive to chlorpyrifos (Fig. 21).

Fewer chronic response data were available for chlorpyrifos. These data were too few for distributional analysis and, because of the short-term nature of exposures, risks from chronic effects were judged minimal. However, these data were used in the calculations of final chronic values. Also, an estimate of the chronic toxicity estimated by the probabilistic analysis was obtained by applying an ACR to the 10th centile of the acute toxicity values.

Additive and Interactive Toxicity. An assumption made in this ecological risk assessment is that chlorpyrifos acts independently of other compounds in the environment. Other organophosphorus insecticides are known to co-occur with chlorpyrifos and, because these act through the same mechanism, would be expected to cause additive toxicity in some locations. In fact, additive toxicity of chlorpyrifos and diazinon has been observed in *Ceriodaphnia dubia* (Bailey et al. 1997). Furthermore, these two pesticides have been observed to co-occur (along with several other organophosphorus pesticides) in surface waters (Mac-Coy et al. 1995). As discussed in the Introduction, the resources to conduct a

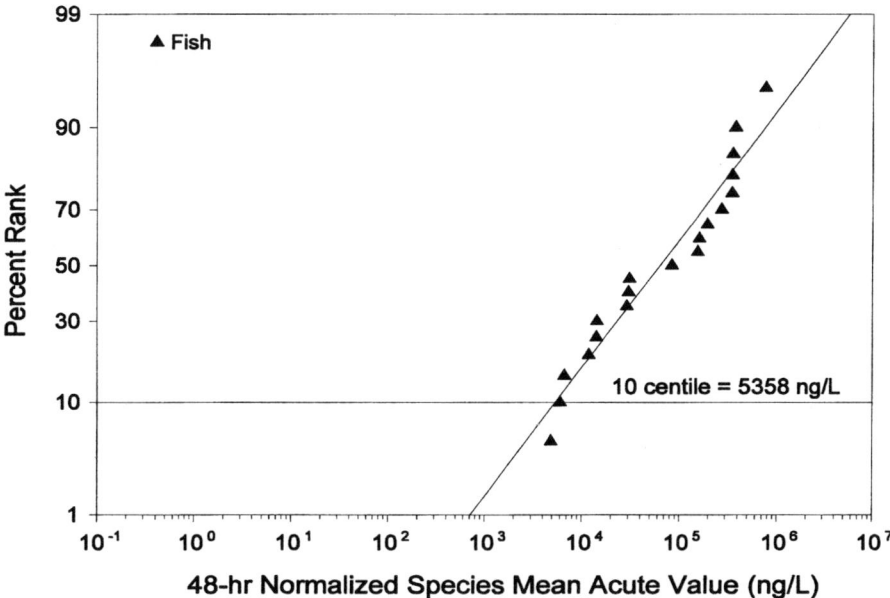

Fig. 22. Distribution of 48-hr normalized EC_{50}/LC_{50} for chlorpyrifos for freshwater vertebrates.

comprehensive risk assessment of all pesticides that may co-occur were not available and, although an additive model could have been applied to those with a common mode of action, these models have not been validated for probabilistic risk assessment purposes.

Recently it has been suggested that other nonorganophosphorus agrochemicals may be synergistic to the toxicity of organophosphorus insecticides to nontarget organisms. Specifically, atrazine, a triazine herbicide that is not considered highly toxic to animals and has a different mechanism of action, has been suggested to synergize the toxicity of chlorpyrifos (Pape-Lunstrom and Lydy 1997). Laboratory studies of binary and tertiary mixtures including chlorpyrifos have indicated that mixtures of the two compounds were marginally more toxic than predicted from a simple additive model of independent action. The study indicated a possible nonadditive response resulting in a supraadditive toxicity (twofold) of combinations of chlorpyrifos and atrazine and found this degree of nonadditivity to be about twofold. When equipotent mixtures were used such that a toxicity of 1.0 toxic unit (TU) was predicted, a value of 0.58 was observed. Several factors, some of which are pointed out by the authors of the report, suggest that this finding is possibly an artifact. First, an artificially small estimate of the LC_{50} for atrazine >20 mg/L was used in the analysis. This assumption was necessary because of the limited water solubility of atrazine relative to the LC_{50}. Second, the studies were conducted at concentrations of atrazine

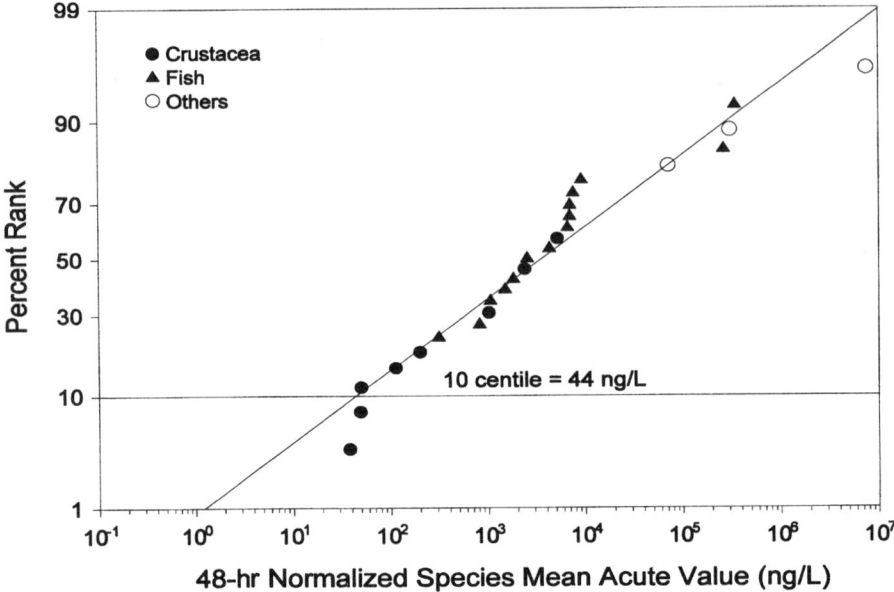

Fig. 23. Distribution of 48-hr normalized EC_{50}/LC_{50} for chlorpyrifos for saltwater organisms tested.

that seldom occur in the environment. Third, for the purposes of their assays, atrazine was dissolved in dimethyl sulfoxide, whereas the other pesticides were dissolved in acetone. It is possible that the uptake of chlorpyrifos was enhanced by the dimethyl sulfoxide added at the same time as the atrazine. Dimethyl sulfoxide is known to enhance uptake of a number of organic substances by aquatic and terrestrial organisms and, ideally, the experiment should have been conducted with the same solvent for all pesticides. The authors report that, at the treatment that would have resulted in 1.0 TU, the concentration of atrazine was 10,000,000 ng/L and that of chlorpyrifos was 380 ng/L. Thus, the ratio of atrazine to chlorpyrifos was 26,315. The ratio of atrazine to chlorpyrifos, based on annual maxima in the Maumee River, OH, for 1987–1995, ranges from approximately 3.0 to 73. The ratio of atrazine tested to determine the synergism was considerably greater than has been observed in the environment. Therefore, the synergism observed is judged to be unlikely to occur in surface waters receiving a combination of these agrochemicals from normal uses.

E. Final Acute and Final Chronic Value Calculation for Chlorpyrifos

As part of the risk assessment, the results of the hazard values calculated from the species probability approach were compared to final acute values (FAVs) and final chronic values (FCVs). These techniques, which are mandated by EPA for use in the Great Lakes Basin, are calculated by the methods to derive water

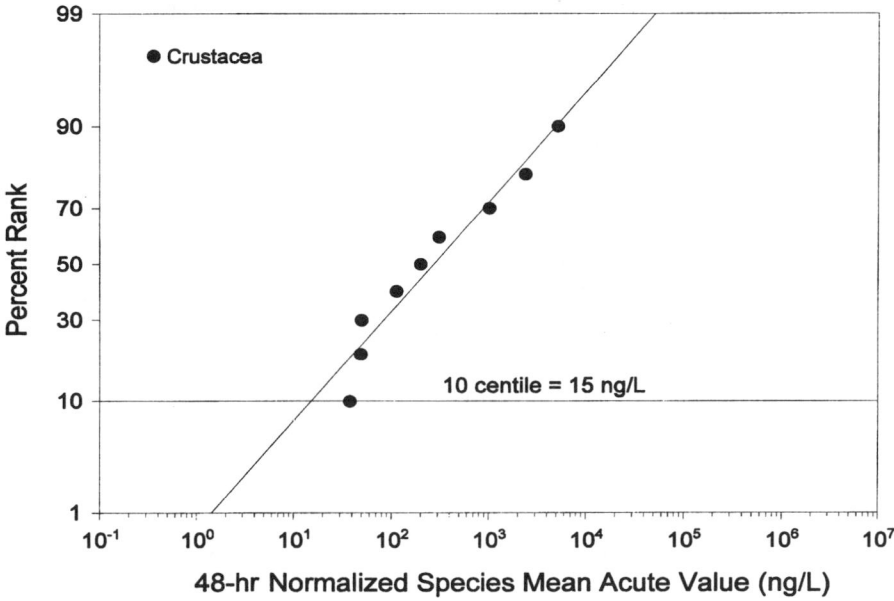

Fig. 24. Distribution of 48-hr normalized EC_{50}/LC_{50} for chlorpyrifos for saltwater arthropods.

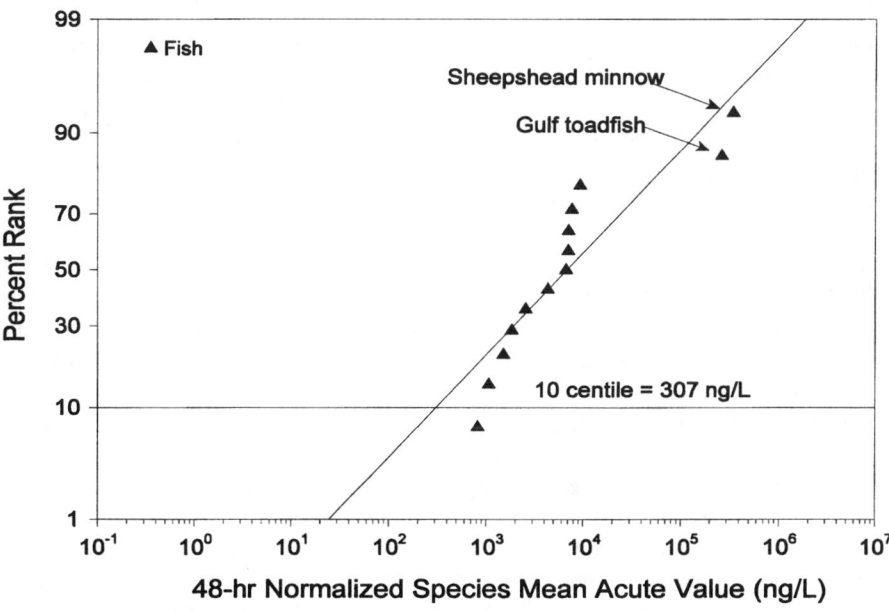

Fig. 25. Distribution of 48-hr normalized EC_{50}/LC_{50} for chlorpyrifos for saltwater fish.

Table 20. Summary of the 10th centiles for the 48-hr LC_{50}/EC_{50} for chlorpyrifos.

Water composition	Organism	EC_{50}/LC_{50} SMAV 10th centile concentration (ng/L)	r^2	N
Freshwater	Arthropods	55	0.955	60
	All organisms[a]	102	0.983	93
	Vertebrates	5358	0.939	19
Saltwater	Anthropods	15	0.94	9
	All organisms[a]	44	0.944	25
	Vertebrates	832	0.919	11

[a]Included snails, rotifers, etc., i.e., nonarthropod invertebrates. All original data taken from Barron and Woodburn (1995) and the 48-hr LC_{50}/EC_{50} estimated from the reciprocity relationships. All data based on EC_{50}/LC_{50} Species Mean Acute Values (SMAVs).

quality standards for the Great Lakes System (USEPA 1995) that make use of a form of distributional (probabilistic) analysis to establish multispecies criteria for hazard assessment. In a manner similar to the probabilistic approaches described previously, these criteria can be compared to distributions of exposures to assess risks.

Final Acute Value (FAV). Studies were chosen to fill the requisite categories for computation of a Tier I FAV by use of selection criteria. The guidelines for the calculation of FAVs (USEPA 1995) suggest that data from acute tests for the calculation of FAVs should include: one test from the class Osteichthyes, the family Salmonidae; class Osteichthyes, other than Salmonidae, preferably an important warm water species; a third representative of the phylum Chordata (an aquatic vertebrate); a planktonic crustacean; a benthic crustacean; an insect; a representative of a phylum other than Chordata or Arthropoda; and a representative of any order of insect or any phylum not already represented. Acute and chronic data are required for at least one fish; one invertebrate; one freshwater species (the other two may be saltwater species); and at least one freshwater alga or vascular plant. Preferred studies were those conducted with freshwater species under flow-through exposure conditions and with the concentration of chlorpyrifos measured. If such experimental data were not available, static-renewal or simple static studies with nominal concentrations were acceptable for the desired species. Where multiple toxicity test data for the same species were available, a species mean acute value (SMAV) was calculated as the geometric mean of the results of all acceptable flow-through acute toxicity tests with the most sensitive tested life stage of the species in which the concentration of the test material was measured. For each genus for which one or more SMAVs were available, the genus mean acute value (GMAV) was calculated as the geometric mean of the SMAVs. The GMAVs were ranked from greatest to least and the cumulative probability, P, for each GMAV calculated from the formula $P = \text{Rank}/(N+1)$ (Table 21). Because there were fewer than 59

GMAVs, as recommended in the Great Lakes Initiative (GLI) procedure, the four least GMAVs were used to calculate the Tier 1 FAV using Eqs. 3–6. Applying the stated criteria to the chlorpyrifos database produced 35 acute studies that met the conservative criteria. The chlorpyrifos acute data set was ranked, sorted, and probabilities assigned to GMAVs, and the product calculated (Table 21). Using the four least GMAVs, a chlorpyrifos FAV of 148 ng/L was calculated (Eq. 6).

$$S^2 = \frac{\Sigma((\ln GMAV)^2) - \frac{(\Sigma(\ln GMAV))^2}{4}}{\Sigma(P) - \frac{(\Sigma(\sqrt{P}))^2}{4}} \quad (3)$$

$$L = \frac{\Sigma(\ln GMAV) - S(\Sigma(\sqrt{P}))}{4} \quad (4)$$

$$A = S(\sqrt{0.05}) + L \quad (5)$$

$$\text{Tier I } FAV = e^A \quad (6)$$

Final Chronic Value (FCV). Using acute lethality data to estimate chronic toxicity to aquatic organisms has generally involved deriving an application factor (AF), termed the acute-to-chronic ratio (ACR), to estimate allowable chronic exposure concentrations for other species for which only acute toxicity data exist (Kenaga 1982). The ACR is derived by dividing the acute LC_{50} concentration by the chronic maximum acceptable toxicant concentration (MATC) for species for which both acute and chronic information are available. It should be noted that one limitation to this approach is that biological endpoints/responses are often not comparable between acute and chronic studies. Acute studies generally use lethality as the endpoint, while a chronic MATC is often derived from an endpoint other than lethality (growth, reproductive ability, etc.). Although the mode of action for lethality is assumed to be the same under acute and chronic exposures, the mode of action may not be the same for different endpoints. For a variety of organisms and chemicals, the ACR has been found to be approximately 10 (Kenaga 1982). This value is supported by quantitative structure–activity relationships (QSARs) for some chemical groups in which acute and chronic regressions are about an order of magnitude apart (Abernethy et al. 1988; Call et al. 1985; McCarty 1986). A range of ACR values was observed for chlorpyrifos. The mean ACR was 8, with a range of 1.4 to 181. Application of the average chlorpyrifos ACR of 8 to the FAV of 148 ng/L gave an FCV of 18 ng/L.

Table 21. Data use for the calculation of the chlorpyrifos final acute value (FAV).

LC_{50}/EC_{50} (ng/L)	Time (d)	Estimated 48-hr LC_{50}/EC_{50} (ng/L)[a]	48-hr SMAV (ng/L)[b]	48-hr GMAV (ng/L)[c]	Reference	Rank	Probability
a. Class Osteichthyes, Family Salmonidae							
Rainbow trout, *Oncorhyncus mykiss* Walbaum							
8,000	4	11,314			Holcombe et al. 1982		
9,000	4	12,728	12,000	12,000	Phipps and Holcombe 1985	12	0.545
b. Class Osteichthyes, other than Salmonidae							
Bluegill sunfish, *Lepomis macrochirus*							
10,000	4	14,142	14,142	14,142	Phipps and Holcombe 1985	13	0.591
Channel catfish *Ictalurus punctatus*							
806,000	4	1,139,856	1,139,856	1,139,856	Phipps and Holcombe 1985	19	0.864
Goldfish, *Carrassius auratus*							
>806,000	4	1,139,856	1,139,856	1,139,856	Phipps and Holcombe 1985	20	0.909
Longnose killifish, *Fundulus similis*							
4,100	4	5,789	5,789	5,789	Schimmel et al. 1983	7	0.318

Species					Reference		
Stickleback, *Gastrosteus aculeatus*							
8,500	4	15,902			van Wijngaarden et al. 1993		
8,500	?ᵃ	6,010	9,776	9,776	van Wijngaarden and Leeuwangh 1993	11	0.5
Stickleback, *Pungitius pungitius*							
4,700	4	6,647	6,647	6,647	van Wijngaarden et al. 1993	8	0.364
c. A third representative of Phylum Chordata							
Fathead minnow, *Pimephales promelas*							
203,000	4	287,085			Holcombe et al. 1982		
140,000	4	197,900			Jarvinen and Tanner 1982		
120,000	4	169,706			Jarvinen and Tanner 1982		
542,000	4	766,504		293,235	Phipps and Holcombe 1985	18	0.818
d. A planktonic crustacean							
Cladoceran, *Daphnia pulex*							
210	2	210	210	210	van Wijngaarden and Leeuwangh 1993	2	0.091
Amphipod, *Gammarus fasciatus*							
180	?ᵃ	127	127		USEPA 1986		
Amphipod, *Gammarus pseudolimnaeus*							
180	?ᵃ	127			USEPA 1986		
300	2	300			Siefert 1984		
200	4	283	221		Siefert 1984		

(*Continued*)

Table 21. Continued.

LC_{50}/EC_{50} (ng/L)	Time (d)	Estimated 48-hr LC_{50}/EC_{50} (ng/L)[a]	48-hr SMAV (ng/L)[b]	48-hr GMAV (ng/L)[c]	Reference	Rank	Probability
Amphipod, *Gammarus pulex*							
70	4	99	99	141	van Wijngaarden and Leeuwangh 1993	1	0.045
e. A benthic crustacean							
Crayfish, *Orconectes immunis*							
6,000	4	8,385	8,485	8,485	Phipps and Holcombe 1985	9	0.409
Isopod, *Proasellus coxalis*							
>20,000	4	28,248	28,248	28,248	van Wijngaarden et al. 1993	14	0.636
f. An insect							
Diptera, *Chaoborus obscuripes*							
6,600	4	9,334	9,334	9,334	van Wijngaarden et al. 1993	10	0.455
Mayfly, *Caenis horaria*							
>3,000	4 (LC_{10})	4,243	4,243	4,243	van Wijngaarden et al. 1993	6	0.273
Mayfly, *Cloen dipterum*							
250	3	400			van Wijngaarden and Leeuwangh 1993		
300	4	367	383	383	van Wijngaarden et al. 1993	4	0.182

Species					Reference		
Mayfly, *Ephemerella* sp.							
400	2			383	Siefert 1984		
300	3		383		Siefert 1984	4	0.182
Water strider, *Gerris gibbifer*							
2,000	4	2,828		2,828	van Wijngaarden et al. 1993	5	0.227
g. A representative of a phylum other than Chordata or Arthropoda							
Mollusca, snail, *Anius vortex*							
>94,000	10	210,190			van Wijngaarden and Leeuwangh 1993		
>94,000	4	132,936	167,158	167,158	van Wijngaarden et al. 1993	15	0.682
Mollusca, snail, *Aplexan hypnorum*							
>806,000	4	1,139,856	1,139,856	1,139,856	Phipps and Holcombe 1985	21	0.955
Mollusca, snail, *Bithynia tentuculata*							
>94,000	10	210,190			van Wijngaarden and Leeuwangh 1993		
>94,000	4	132,936	167,158	167,158	van Wijngaarden et al. 1993	16	0.727
Mollusca, snail, *Lymnaea stagnalis*							
>94,000	10	210,190			van Wijngaarden and Leeuwangh 1993		
>94,000	4	132,936	167,158	167,158	van Wijngaarden et al. 1993	17	0.773

[a] Estimated 48-hr LC_{50}/EC_{50}s (ng/L) were calculated.
[b] SMAV, Species Mean Acute Value (geometric mean of acceptable LC_{50}/EC_{50} values).
[c] GMAV, Genus Mean Acute Value (geometric mean of acceptable SMAV for organisms in that genus).
[d] Worst-case assumption made on time (d) = 1 day.

F. Toxicity of Sediment-Borne Chlorpyrifos

Chlorpyrifos is moderately hydrophobic and tends to partition into organic matter. Therefore, chlorpyrifos can be accumulated in detritus and sediment. Fugacity theory predicts that sediment-dwelling organisms would accumulate chlorpyrifos to a concentration proportional to the concentration of chlorpyrifos in pore water (interstitial water), which would be indirectly proportional to the concentration in the bulk sediment, the lipid content of the organism, and the organic carbon-normalized partitioning coefficient (K_{OC}) and inversely proportional to the concentrations of organic carbon in the bulk sediments and pore water (Mansingh et al. 1997). Chlorpyrifos is moderately accumulated by invertebrates, to an extent similar to that by fish (Montañés et al. 1995). The bioconcentration factors for the freshwater isopod *Asellus aquaticus* ranged from 262 ± 77 to $235 \pm 44 \times 10^3$ L·kg^{-1} (normalized for lipid content), for two different water concentrations, respectively (Montañés et al. 1995). These theoretical considerations are the basis of the equilibrium partitioning method of deriving sediment quality criteria for the protection of benthic organisms from the effects of sediment-bound toxicants (DiToro et al. 1991; Hoke et al. 1994).

If chlorpyrifos is partitioned from the water column to sediment, protection of organisms dwelling in the water column should be expected to protect sediment-dwelling organisms. The logic for this conclusion is as follows: chlorpyrifos accumulation into sediments is a dynamic process. There is degradation of chlorpyrifos in sediments and, when the concentration gradient is from the sediment to the water, a combination of degradation and diffusion back into the water column would be expected to decrease concentrations in the sediment. Accumulation of chlorpyrifos directly from the water column into sediments would be proportional to the K_{OC} of the compound. Toxicity of chlorpyrifos to animals in the water column is dependent on the concentration of chlorpyrifos in the water column, and toxicity of chlorpyrifos in the sediment would be proportional to the concentration of chlorpyrifos in the interstitial water. Because at steady state the fugacities of chlorpyrifos would be equal in the water column and pore water, it can be concluded that the relative toxicities of the water column and sediment pore water would be equal at steady state and that the relative toxicities of chlorpyrifos in the pore water and water column under non-steady-state conditions would be proportional to the rates of movement between the sediment pore water and water column. Because the release from runoff to surface water is relatively rapid and of short duration and the diffusion-limited process of transfer back into the water column from sediment is relatively slow, it is logical to conclude that it would be impossible for the concentration of chlorpyrifos in the pore water to exceed the time-weighted average of chlorpyrifos in the water column. Given that animals in the water column have tolerances of chlorpyrifos similar to those of benthic invertebrates, if organisms in the water column are protected, it is unlikely that there would be accumulations of chlorpyrifos into the sediments that would cause adverse effects to benthic organisms.

In addition to direct partitioning of chlorpyrifos to the sediment from the water column, it is possible for chlorpyrifos-containing sediment or detritus to be deposited directly into the sediments (Bergamaschi et al. 1997). In this case, a disequilibrium with the water column could occur and the pore water concentration of chlorpyrifos could exceed the time-weighted average in the water column (Domagalski and Kuivila 1993). Therefore, an assessment of the likelihood of sediment concentrations of chlorpyrifos to exceed a critical concentration from deposition of chlorpyrifos on suspended sediments was performed.

Little empirical information on the toxicity of sediment-bound chlorpyrifos to freshwater benthic invertebrates is reported in the literature. The EC_{50} value for toxicity of chlorpyrifos to the midge *Chironomus tentans* was determined to be 383 (95% CI = 358–409) µg/kg dry weight (dw), while that for *Hyalella azteca* was 399 (95% CI = 355–448) µg/kg dw (Brown et al. 1997). The toxicities of chlorpyrifos to *C. tentans* when spiked into sediments containing 3.0% and 8.5% organic carbon were found to be 522 and 424 µg/kg dw, respectively (Ankley et al. 1994). The mean EC_{50} from the two locations studied by Ankley et al. (1994) was 69.5 µg/L, which is similar to the water-only value of 70 µg/L (Brown et al. 1997). This indicates that the equilibrium partitioning (EqP) approach is a useful predictive tool for estimating toxicity of chlorpyrifos in sediments. Therefore, it is appropriate to use the acute-to-chronic ratio of 8.0 derived from water-only exposures (Table 22) to estimate the chronic NOAEC for sediment from the LC_{50} values. When this is done, NOAEC values for *C. tentans* and *H. azteca* of 48 and 50 µg/kg dw, respectively, are estimated. If the FCVs of 20 or 70 µg/L and the USEPA EqP methodology are used to estimate the sediment quality criteria for 3% or 8% organic carbon content, the sediment quality criteria (SQC) range from 2.3 to 27.7 µg/kg dw (Table 23). From the data of Brown et al. (1997), ranges of probable effects can be determined (Table 24).

Only one study of the toxicity of chlorpyrifos in sediment was found for marine benthic organisms. In that study, the effects of chlorpyrifos on population-level responses of the benthic marine copepod *Amphiascus tenuiremis* were determined. It was concluded that there were significant population-level effects from exposure to 6 µg/kg, dw (Green and Chandler 1996). There was slightly less survival of adult copepods as well as reductions in intrinsic rate of natural increase (r). These effects were observed at concentrations that were approximately 7%–36% of the species LC_{50} value, which would be equivalent to acute-to-chronic ratios from 3 to 14. However, there was no concentration–response relationship for these data so that it was impossible to obtain an accurate estimate of the reference concentration for the effects of chlorpyrifos on this species. As it is unlikely that concentration of chlorpyrifos in marine sediments will exceed those in freshwater, we do not suggest that additional information be gathered on toxicity of chlorpyrifos to marine benthic organisms. A calculation analogous to that presented in Table 23 could be conducted for saltwater.

When chlorpyrifos has been added to enclosures or mesocosms, it has been found to rapidly dissipate from the water column, and some fraction becomes

Table 22. Data for calculation of acute to chronic ratios for chlorpyrifos.

Species	Acute LC_{50}/EC_{50} (ng/L)	Chronic MATC (ng/L)	Acute to chronic ratio (ACR)	Study
Freshwater				
Daphnia magna	1,700	75	23	Acute = static-renewal, measured and not measured
Fathead minnow	293,000	1,619	181	Acute = geometric mean of 4 studies, flow-through, measured, chronic = geometric mean of 2 embryo-larval studies and 1 life-cycle study
Saltwater				
Mysid shrimp	35	7	5	Acute = flow-through, not measured, chronic = flow-through, measured
Atlantic silverside	1,297	370	3.5	Acute = flow-through, measured
California grunion	1,068	500	2.1	Acute = flow-through, measured
Gulf toadfish	68,000	2,280	30	Acute = flow-through, measured
Inland silverside	4,200	1,160	3.6	Acute = flow-through, measured
Tidewater silverside	753	540	1.4	Acute = flow-through, measured
		Overall ACR:	8	

MATC, Maximum Acceptable Toxicant Concentration; ACR, Acute to Chronic Ratio

Table 23. Sediment quality criteria using equilibrium partitioning (EqP) theory and assuming a log K_{OC} of 3.92 and 3% or 8% organic carbon contents of sediments. Chronic toxicity values of 20 or 70 ng/L were assumed.

FCV (µg/L)	OC (%)	Freshwater criterion (µg/kg OC)	Freshwater criterion (µg/kg sediment)	Upper 95% CI (µg/kg sediment)	Lower 95% CI (µg/kg sediment)
0.02	3	166	2.3	5	10.7
0.02	8	166	6.2	15.3	28.6
0.07	3	52	8.1	17.5	37.5
0.02	8	52	21.7	46.6	100

OC, organic carbon.

associated with the sediment. However, relatively little chlorpyrifos accumulated in the sediments (Crum and Brock 1994; Giddings et al, in manuscript; Leeuwangh 1994). When 5 × 10 m enclosures were treated with 20 µg/L, 95% of the chlorpyrifos had dissipated from the water column in 7 d (Knuth and Heinis 1992). After 1 d, 3.2% was in the sediment and, after 64 d, 0.17% was in the sediment. By 420 d post treatment, 0.5% of the chlorpyrifos originally added was in the sediment. In a similar study conducted in 11-m^3 aquatic microcosms, concentrations of chlorpyrifos both dissolved in water and adsorbed to particulates in a slurry were added at rates that bracketed maximum likelihood runoff scenarios. These two treatments resulted in concentrations of chlorpyrifos of 1–275 µg/kg (dw) sediment in the aqueous-treated and 1–93 µg/kg (dw) in the slurry-treated microcosms (Giddings 1993a,b; Giddings et al., 1997).

Relatively few measurements of chlorpyrifos in sediments have been reported. In the San Joaquin River of the Central Valley of California near Vernalis, concentrations of chlorpyrifos ranged from 8.6 to 30 µg/kg, dw, of sediment (Bergamaschi et al. 1997). In another study of the San Joaquin River, concentrations of chlorpyrifos in suspended sediments ranged from less than detection (<0.5 ng/L) to as much as 153 ng/L (Pereira et al. 1996). Because no measures of the total mass of particulates in the water were given, it was impossible to determine the concentration of chlorpyrifos associated with the sediment. Even in the Central Valley of California where concentrations of chlorpyr-

Table 24. Concentration ranges for toxicity of sediments.

Whole-sediment chlorpyrifos concentration	Potential for adverse effect
Less than 100 µg/kg	Not probable
Between 100 and 500 µg/kg	Possible
Greater than 500 µg/kg	Probable

Source: Brown et al. (1997).

ifos were determined to be sufficiently great to adversely affect water column arthropods, the concentration in Dry Creek (153 ng/L) was the only measurement that might be toxicologically relevant.

When the concentrations of chlorpyrifos reported to occur in freshwater sediments from field or mesocosm studies were compared to the toxicity values for the midge and amphipod, there were no exceedences for most situations except for the Central Valley of California. Thus, it can be concluded that, in the sites studied, little risk is posed by the concentrations of chlorpyrifos observed in sediments to benthic invertebrates except for some sites in the San Joaquin River, where water concentrations were also deemed to be potentially toxic to water column arthropods. When toxicity values for the 60 arthropod species included in the data set used to derive the hazard values are normalized to 10-d exposures, using the equations presented in the section on reciprocity, the values for the midge and amphipod are among the least tolerant species (ranks 4 and 6, respectively). Thus, conclusions based on these two species are probably protective of most other benthic invertebrates.

No information on concentrations of chlorpyrifos in marine sediments was available for comparison to the reference dose for marine copepods. Thus, for this type of assessment, a research need would be to make more measurements of chlorpyrifos concentrations in marine and estuarine sediments. Effects of sediment-bound chlorpyrifos on estuarine invertebrates have been reported (Chandler et al. 1997). The study was conducted to determine the effects of sediment-bound chlorpyrifos on intact estuarine sediments. Concentrations of 21–33 ng chlorpyrifos/g, dw, were studied. None of the concentrations tested caused any statistically significant effects on the total meiobenthos population densities. The predominant naturally occurring copepod, *Microarthridion littorale,* was decreased by some of the chloryrifos-spiked sediments, but all the other copepods were either unaffected or had their populations enhanced by the chlorpyrifos treatments. These authors indicated that the concentrations of chlorpyrifos used in their studies were approximately 10 fold less than the concentration of 245 ng chlorpyrifos/g, dw, which was the greatest concentration reported for U.S. estuaries (Pait et al. 1992). This information does not indicate that there is significant risk from current concentrations of chlorpyrifos in estuarine sediments. To reduce uncertainty in this conclusion would require additional information on the concentrations of chlorpyrifos in estuarine sediments.

G. Microcosm and Mesocosm Studies with Chlorpyrifos

Microcosms and mesocosms are small-and medium-scale experimental ecosystems, often used to study the fate and effects of chemicals under quasi-natural conditions. These systems can provide valuable information that cannot be easily obtained from simpler studies. Specifically, (1) microcosms and mesocosms contain diverse communities of microorganisms, plants, zooplankton, and benthic invertebrates, allowing simultaneous measurement of chemical effects on a wide variety of taxonomic groups; (2) microcosms and mesocosms incorporate interactions among populations, such as predation and competition, and therefore allow observation of indirect ecological impacts of chemicals (for example, reduction in growth of one species because of removal of that species food supply by a toxic chemical); and (3) microcosms and mesocosms incorporate mechanisms of ecological recovery from chemical stress (such as replacement of chemical-sensitive species by more tolerant species), although to a lesser extent than in open natural ecosystems. In microcosm and mesocosm studies, chemical exposure regimes (including frequency and duration of exposure, partitioning between sediment and water, abiotic dissipation pathways, and additional inputs such as solar insolation) can be more realistic than in a typical laboratory study.

Microcosm and mesocosm studies with chlorpyrifos have been reviewed by Leeuwangh (1994) and Barron and Woodburn (1995). Most of the information comes from four groups of researchers: Hurlbert and co-workers at the University of California, Riverside (Hurlbert et al. 1970, 1972); the Duluth laboratory of the U.S. Environmental Protection Agency (Brazner and Kline 1990; Siefert et al. 1987, 1989); Springborn Laboratories' studies on behalf of Dow AgroSciences (Biever et al. 1994; Giddings 1993a, b; Giddings et al. 1997); and an extensive series of studies at the Winand Staring Center in the Netherlands (Brock et al. 1992, 1995; Crum and Brock 1994; Cuppen et al. 1995; Leeuwangh 1994; Leeuwangh et al. 1994; Lucassen and Leeuwangh 1994; van den Brink et al. 1995, 1996; van Donk et al. 1995; van Wijngaarden et al. 1996). All these studies used emulsifiable concentrate formulations of chlorpyrifos, applied as diluted sprays or, in the case of some of the Springborn treatments, mixed with soil and applied as a slurry.

Pond Studies at the University of California, Riverside. Experimental ponds at the University of California, Riverside, pond facilities were sprayed four times at 2-wk intervals with chlorpyrifos (Dursban®), at rates of 0.011, 0.056, 0.11, and 1.1 kg a.i./ha (Hurlbert et al. 1970). Chlorpyrifos concentrations measured in the water 4 hr after application were 200,000 ng/L at the greatest application rate and 10,000 ng/L at the second least application rate; if the same relationship applied to the other rates, the least application rate corresponded to approximately 2,000 ng/L chlorpyrifos. Populations of the dominant zooplankton species, *Cyclops vernalis* (a copepod) and *Moina micrura* (a cladoceran), were reduced by the chlorpyrifos treatments, as was that of the corixid *Corisella*. Populations of the copepod *Diaptomus pallidus* and the rotifer *Asplanchna brightwelli* increased in the ponds where *Cyclops vernalis* and *M. micrura* populations were reduced (except for *D. pallidus* at the greatest application rate). Survival and reproduction of mosquitofish (*Gambusia affinis*) were unaffected.

To investigate the ecological relationships that led to increases in numbers of *D. pallidus* and *A. brightwelli*, the study was repeated using chlorpyrifos application rates of 0.28 and 0.028 kg a.i./ha (72,000 and 7,200 ng/L, respectively), repeated three times at 2-wk intervals (Hurlbert et al. 1972). The responses of *C. vernalis, M. micrura, D. pallidus,* and *A. brightwelli* were the same as in the first experiment. *C. vernalis* and *M. micrura* recovered in 1–3 wk at the lesser application rate and in 3–6 wk at the greater rate. Besides *D. pallidus* and *A. brightwelli*, populations of herbivorous rotifers increased soon after *C. vernalis* and *M. micrura* decreased. An algal bloom developed that was attributed to the loss of herbivorous crustaceans; 6 wk after treatment, algal densities in ponds at the lesser application rate were twice as great as controls, and those at the greater application rate were 16 times as great as those of the controls. Among the insects, populations of the predaceous hemiptera (notonectids and corixids) declined after chlorpyrifos treatment and recovered slowly, while the herbivorous insects were less affected and recovered more quickly.

Littoral Enclosure Studies at the Duluth EPA Laboratory. In the study conducted at the EPA mesocosm facility, littoral enclosures in a pond were sprayed once with chlorpyrifos at rates of 0.0029, 0.032, and 0.123 kg a.i./ha, which resulted in initial chlorpyrifos concentrations of 500, 5,000, and 20,000 ng/L, respectively. Populations of the ostracod *Cyclocypris* and all five cladoceran species present in the enclosures were significantly fewer at all application rates 4 d after treatment (Brazner and Kline 1990; Siefert et al. 1989). Copepod densities were less, on average, in the treated enclosures than in the controls, but few of the differences were statistically significant. Populations of a few rotifer species were reduced at the least application rate, but most were unaffected at the greater application rates, and some were more abundant in the treated enclosures, on average, than in the controls. Populations of most species of chironomids (the dominant insect group) were significantly reduced at all treatment levels 4 d after treatment with chlorpyrifos. Populations of chironomids recovered to pretreatment levels within 16 d at the least treatment level, but were still less than those in the control enclosures after 32 d in the greater treatment levels. Populations of other insects and the amphipod *Hyalella azteca* were also affected, while snails, planaria, and protozoa were unaffected or increased in the treated enclosures. An algal bloom occurred in the treated enclosures 8 d after chlorpyrifos application.

Survival of juvenile fathead minnows (*Pimephales promelas*) exposed in cages in the enclosures was unaffected by any of the chlorpyrifos treatments, while survival of bluegill sunfish (*Lepomis macrochirus*) was less in the intermediate and greatest treatment concentrations. The difference in response of the two fish species was consistent with their observed sensitivity in laboratory studies. Fathead minnow larvae that hatched in the enclosures 1 wk before treatment grew more slowly than those in the control enclosures (Brazner and Kline 1990). The number and diversity of food items in the stomachs of these fish were less than in the control enclosures at all treatment regimes. The reduction in stomach contents was attributed to a reduction in abundance of invertebrate prey in the enclosures (Brazner and Kline 1990). The reduced larval growth rate was interpreted as an indirect effect of the chlorpyrifos treatment. However, analysis of food selectivity indicated that fathead minnow larvae in the enclosures (including controls, and all enclosures before treatment) preferred rotifers and protozoans to cladocerans and copepods. Rotifers and protozoans were generally unaffected by chlorpyrifos treatment, casting some doubt on the hypothesis that reduction in fish growth was caused by reduced food availability. Thus, it is likely that other factors may have been partly or completely responsible for the observed effects.

Pond Microcosm Studies by Springborn Laboratories. The Springborn studies (Biever et al. 1994; Giddings 1993a,b; Giddings et al. 1997) were conducted in large outdoor pond microcosms near Lawrence, KS. Chlorpyrifos was applied as a surface spray, a slurry of clay particles, or a combination of the two. Concentrations in water immediately after treatment ranged from 30 to 3000 ng/L; 96-hr maxima ranged from 20 to 1860 ng/L. A single spray application resulting in a 96-hr concentration of 120 ng/L reduced the abundance of cladocerans, copepod nauplii (but not adult *Diaptomus pallidus*), mayflies, and chironomid midges. These effects were temporary. Copepod and cladoceran population densities returned to control levels within 2 wk, midges within 4 wk, and mayflies within 8 wk. The abundance of rotifers increased. Growth and survival of juvenile bluegill sunfish were affected at concentrations >1000 ng/L but not less. At greater treatment regimes, effects were more widespread and more pronounced, and recovery took longer. Most rotifer species and some midges (tribe Tanytarsini) were not reduced at any treatment level. Bluegill sunfish growth was affected at 440 ng/L, and survival was reduced at 1410 ng/L.

Chlorpyrifos applied to the microcosms as three biweekly clay slurry treatments produced similar results. Copepod nauplii (not adult *Diaptomus pallidus*) were the most sensitive, with brief reductions (d 1 post treatment) in their populations at exposures as small as 30 ng/L (96-hr maxima) and long-term reductions at nominal concentrations ≥300 ng/L. Cladocerans, ostracods, and some midges were affected at 210 ng/L. Other midges, as well as mayflies, were reduced by a concentration of 1860 ng/L. Bluegill sunfish growth was reduced by 540 ng/L and survival by 1860 ng/L. Most rotifer species and several groups of midges were unaffected at any treatment level. As in the study with spray treatments, most effects at the lesser treatment levels were transient. Copepod nauplii were an exception; at 210 ng/L and greater concentrations, nauplii densities were reduced through the end of the season.

Microcosm and Mesocosm Studies in the Netherlands. Researchers at the Winand Staring Center at Wageningen, the Netherlands, investigated the fate and effects of chlorpyrifos in an extensive series of experiments. Most of the experiments were conducted in 600-L indoor microcosms. Others took place in shallow-ditch mesocosms simulating one of the most characteristic aquatic habitats of that country. The results of all studies were fairly consistent, and together they present a richly detailed picture of the responses of pond ecosystems to chlorpyrifos exposure.

Most of the studies with indoor microcosms involved single spray applications of chlorpyrifos emulsifiable concentrate formulation at initial concentrations of 35,000 ng/L. Results of the most comprehensive of these experiments were summarized by Cuppen et al. (1995). Chlorpyrifos exposure resulted in direct toxicity to cladocerans, copepods, amphipods, isopods, and insects. The reduction in herbivores resulted in greater growth of periphyton and phytoplankton. As the concentration declined, the populations of many herbivores, including copepods, rotifers, snails, and oligochaetes, increased in response to the increased food supply and decreased competition from other, more sensitive herbivores. The reduction in invertebrate shredders (amphipods and isopods) resulted in decreased rates of litter decomposition.

Ecological responses to 35,000 ng/L chlorpyrifos have little relevance to exposures resulting from agricultural applications. However, the Dutch researchers evaluated the effects of lesser chlorpyrifos concentrations in two ways: first, by observing the recovery of populations within the microcosms, and second, by incubating organisms in cages in the microcosms at various times (Leeuwangh et al. 1994). It was found that copepods and some cladoceran populations recovered when chlorpyrifos concentrations decreased to 200 ng/L; populations of other cladocerans recovered when chlorpyrifos concentrations decreased to less than 100 ng/L. Taxa with no source for colonization because they had no resistant life stages, no refugia within the system, and no access for recolonization from outside, such as insects, amphipods, and isopods, did not recover. However, the potential for these populations to become reestablished was evaluated on the basis of the results of the caged exposures. In this way, the no-observed-effect concentrations (NOECs) for *Chaoborus obscuripes* (midge), *Cloeon dipterum* (odonate), and *Gammarus pulex* (amphipod) were determined to be 200 ng/L, and that of *Asellus aquaticus* (isopod) was estimated to be 1,300 ng/L.

Effects of small chlorpyrifos concentrations were determined more directly in an experiment with the ditch mesocosms (van den Brink et al. 1996; van Wijngaarden et al. 1996). The mesocosms were sprayed once with chlorpyrifos at initial concentrations of 100, 900, 6,000, and 44,000 ng/L. The EC_{10}, EC_{50}, and no-effect concentrations for individual taxa were calculated based on the initial 48-hr mean concentrations in mesocosm water. *Chaoborus obscuripes*, *Cloeon dipterum*, and *Gammarus pulex* and *Asellus aquaticus* were exposed in cages in the microcosm experiment described. For most arthropod species, the EC_{50} values ranged from 300 to 600 ng/L. Overall, "a no-observed-effect concentration of 100 ng/L could be derived both at the species and the community level" (van den Brink et al. 1996).

Most taxa affected by greater concentrations of chlorpyrifos in ditch mesocosms recovered before the end of the 24-wk study. Crustacea (except *G. pulex*) recovered rapidly even at the greatest treatment level because of rapid reproduction rates and the presence of resistant resting stages. The amphipod *G. pulex* did not recover because there was no source for recolonization. This result was "an artifact due to the isolated position of the mesocosms;" noting that recolonization sources normally exist in natural ecosystems (van den Brink et al. 1996).

In contrast to the indoor microcosm experiments, few indirect effects on nonarthropods were observed in the mesocosm studies. The authors attributed this to the greater complexity of the mesocosms than the microcosms. "A structurally more diverse and complex ecosystem includes more redundant populations and ... feedback mechanisms, so indirect effects are harder to detect" (van den Brink et al. 1996).

V. Risk Characterization
A. Introduction

As discussed in the problem formulation section, the process of ecotoxicological risk assessment is evolving and many improvements and new approaches have been developed in the last few years (Environment Canada 1996; NRC 1993; USEPA 1992a, 1996a). Currently, the USEPA has a workgroup addressing the development of procedures for probabilistic risk assessment for pesticides (ECOFRAM 1997). These suggested procedures all rely on a basic framework where, in the process of risk characterization, exposure and effects profiles are brought together and integrated into a risk estimate (as in USEPA 1992a). Recent approaches that have been applied to this framework are the use of tiered steps to rationalize resource allocation to the risk assessment processes (SETAC 1994) and the use of probabilistic or distributional procedures for characterizing effects and exposure data.

Tiers in Risk Assessment. The report of the Aquatic Risk Assessment Dialogue Group (ARADG; SETAC 1994), suggested that four tiers be used in the risk assessment process for pesticides in aquatic ecosystems. These tiers began with a simple "worst-case" estimation of environmental concentration, which was compared with the effect level for the most sensitive organism (the hazard quotient approach). If this hazard quotient suggested a potential risk, further tiers of risk assessment with more realistic and more complete exposure and effects data could be applied to the problem. These simple "worst-case" or "one-tailed" assessments are configured as conservative screening tests designed to allow elimination of nonhazardous compounds from further considerations. Because of their nature, this is seldom the case for pesticides, especially insecticides. The fact that a pesticide fails a screening test does not indicate that there would be an unacceptable risk in the environment, but rather that there is a level of concern that warrants expenditure of resources to conduct a more refined risk assessment. The ARADG (SETAC 1994) suggested that the intermediate tiers (II and III) of the assessment process make use of probabilistic approaches while the highest tier (IV) could include specially designed toxicity tests (microcosms, field tests, etc.) as well as assessments based on landscape models. This assessment of the risks of chlorpyrifos in aquatic environments follows a modification of the tiered approach from SETAC (1994). The major modification of the process in the report (SETAC 1994) is the use of probabilities of exceeding measured concentrations of chlorpyrifos in surface water in addition to refined simulation models to estimate environmental concentrations.

Quotients. The first tier in risk characterization process is the use of hazard quotients. These are simple ratios of exposure and effects and may be used to express hazard or relative safety (Eq. 7).

$$Hazard = \frac{Exposure\ conc.}{Effect\ conc.} \quad or: Margin\ of\ Safety = \frac{Effect\ conc.}{Exposure\ conc.} \quad (7)$$

Hazard quotients (HQ) traditionally have been calculated by utilizing the susceptibility of the most sensitive organism or group of organisms and comparing this to the greatest exposure concentration to estimate hazard. The assessment can be made more conservative by the use of safety factors to allow for unquantified uncertainty in the effect and exposure estimations or measurements. The quotient model for ecological risk assessment fails to consider ranges of sensitivity in different species in the ecosystem or ranges of exposure concentration. Expressing the results of an exposure or effects characterization as a distribution of values in a probabilistic approach rather than a single point estimate has the advantage of using all relevant single species toxicity data and, when combined with exposure distributions, allows for a quantitative expression of risks to communities of organisms (Health Council of the Netherlands 1993; SETAC 1994). Techniques for the use of probabilistic risk assessment have been discussed recently (Anderson and Yuhas 1996; Burmaster 1996; Carrington 1996; Power and McCarty 1996; Richardson 1996).

Probabilistic Approaches. Probabilistic approaches to assessing and managing risk have been used in society for many years. The probability of occurrence of an adverse event has been widely used in the characterization of risk from many events in human society (the insurance industry) and for protection against failure in engineering projects (Solomon 1996). This concept has now been applied in ecological risk assessment for the characterization of both exposures and effects but with some caveats. These caveats relate to important differences between traditional ecotoxicological risk assessment and the use of probabilistic approaches and the nature of the assessment endpoints, such as ecosystem structure or function (Solomon 1996). The principle of the probabilistic approach in ecotoxicological risk assessment has been described (Cardwell et al. 1993; Klaine et al. 1996; Parkhurst et al. 1995; SETAC 1994; Solomon et al. 1996) and is illustrated diagrammatically in Fig. 26. It is known that many parameters and measures are distributed in a consistent manner (following a normal or similar distribution) and that, from these distributions, and given a similar population of events, it is possible to assign probabilities to a range of values (i.e., the likelihood that a measure will exceed a certain value). This applies to concentrations of substances in the environment; however, in this case, data are often censored by the limits of analytical detection (Fig. 26A) and they are frequently log-normally distributed. When plotted as a cumulative frequency distribution using a probability scale on the y-axis as a function of $\log_{10}$ concentration (Fig. 26B), these distributions approximate a straight line that can be

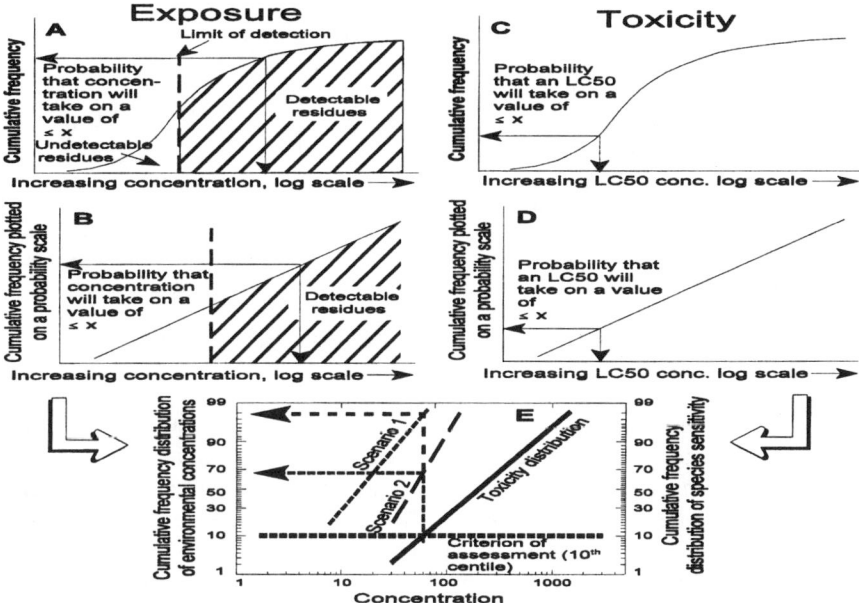

Fig. 26. A–E. Illustration of the principle of the probabilistic approach (adapted from Solomon and Chappel 1997).

used to estimate the likelihood that a particular concentration of the substance will be exceeded in an environment where the circumstances controlling the fate and concentration of the substance are similar to the sampled environment. A similar approach can be taken with susceptibility of different organisms in the ecosystem to the substance (Fig. 26C,D). The use of these distributions in the probabilistic characterization of risk is illustrated in Fig. 26E. In this procedure, it is assumed that the distributions of sensitivity represent the range of responses that are likely to be encountered in the ecosystems where the exposures occur (SETAC 1994). If the exposure data were collected over time at a particular site, the degree of overlap of the exposure distribution with the effects distribution can be used to estimate the joint probability of exposure and toxicity, leading to estimates of exceedence probabilities for responses at a fixed effect assessment criterion, such as, for example, the concentration equivalent to the 10th centile of the species distribution (Fig. 26C). This can be applied to a number of data sets (scenarios 1 and 2, Fig. 26E) and the resulting probabilities used for priority setting or in further assessing ecological relevance. The relationship between the distribution of species sensitivity and the exposure distribution can be extended by calculating the exposure exceedence probabilities for a number of point estimates from the species distribution. This can be presented graphically as an exceedence profile plot for ease of interpretation (Fig. 27).

The ecological risk assessment process becomes limited when attempts are

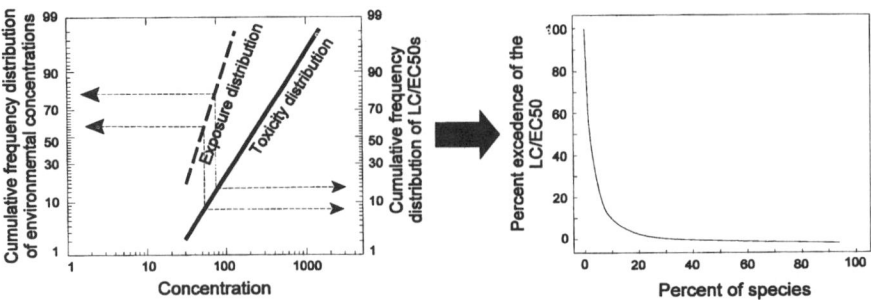

Fig. 27. Illustration of the derivation of an exceedence profile plot.

made to predict absolute "no-effect" exposures. When using absolute no-effect levels, great changes in exposure result in marginal changes in incremental risk. This can result in overestimates of risk or unrealistically small allowable exposures for a predicted level of risk. It is difficult to discriminate between changes in response when on the plateau regions of the sigmoid dose–response relationship. In a probabilistic sense, there are no zero or 100% responses. The nearer to the extremes the more difficult it is to accurately discriminate changes in response as a function of exposure. For this reason, estimates of any response other than zero or 100% are more accurate and useful in risk assessments. In an attempt to add assurances to safety, uncertainty factors (safety factors) to account for uncertainties in the risk assessment process are frequently used. Thus, the use of these types of approaches to assure protection often results in unrealistically low levels of allowable exposure. A better approach is the use of a reference (benchmark) dose based on responses that are measurable with the type of toxicity data that are usually available (USEPA 1995b). These responses can be calibrated with population-, community-, or ecosystem-level effects.

The use of probabilistic approaches allows for responses to be quantified in a meaningful way. The expression of response on a probabilistic basis also facilitates risk communication. The probabilistic approach allows quantification of likelihood of effects, which, by definition, is risk. This cannot be done by the use of simple hazard quotients. The probabilistic approach is useful even for simple hazard quotients in that it examines more of the universe of possible exposures and responses and quantifies the probability of observing a more extreme value for either exposure or hazard. For instance, by examining the entire range of tolerances of species, even though the response of the most sensitive species may be used to estimate a reference dose, information on the probability of finding a more sensitive species that might occur in the environment of interest would be available.

The probabilistic approach is flexible and allows the separation of risk assessment from risk management. Once the probabilistic relationships have been established, socially relevant assessment endpoints can be applied. The level of protection or probability of adverse effect can be selected and applied to the same distributions without the need to completely conduct the entire risk assessment again. This is useful in the risk–benefit phase of the analysis. That is, risk management decisions can easily be separated from the risk assessment phase. For instance, allowable concentrations that would result in probabilities of exceedence of a range of centiles from the toxicity distributions could be determined from the exposure and hazard functions. The decision of which to adopt could be based on other considerations related to costs and benefits, but the analysis would not need to be repeated. In short, the probabilistic methodology gives more information to the risk manager than does the simple hazard quotient method. This technique also avoids the miscommunication fostered when individuals think that the hazard quotient values represent absolute standards for environmental protection. The probabilistic risk assessment process is more transparent than the hazard quotient methodology, which makes it easier for the risk communicator to communicate that, in fact, there is a continuum of both exposure and hazard to be considered. The assumptions made do not need to be as extreme to allow protection of the resource at risk. The probabilistic assessment procedure because of its flexible nature is less prone to stakeholders taking extreme positions that are more difficult to reconcile with resources and social needs.

When data are summarized or statistical analyses applied, the resulting measures are expressed as rates that are, by definition, population-level parameters. For instance, mortality is a rate that has no relevance for individuals. Individuals can be dead or alive, which is a quantal response. Individual organisms do not have a mortality rate. This is an aggregate function that is a population-level property. The smallest experimental units for assessment and measurement endpoints are individual organisms of individual species. It is not possible to estimate whether an adverse effect will occur to a particular individual organism as its position on the concentration–response curve is not known *a priori*. However, protection of the population, its sustainability, or its function in the ecosystem is more properly used as an assessment endpoint in ecotoxicological risk assessment. The concentration–response curve allows the estimation of the likelihood that a certain proportion of the population of a species will be affected at a given exposure. This can be used for risk assessment in individual populations in a probabilistic sense, particularly if it is known what rates of removal of organisms from the population can be sustained. The probabilistic risk approach allows the more sensitive species to be easily identified and the probability of observing a more sensitive species to be estimated. Furthermore, the process allows the calculation of the probability of exceedence between the sensitivity of appropriate species and their exposures. These species represent surrogates for other potentially sensitive species. For instance, in this probabilistic risk assessment of chlorpyrifos, the 10th centile of the distribution of acute toxicities to species was used as the assessment endpoint. Species representing those with probabilities of occurrence of less than below the 10th centile included mosquito larvae that do not occur in running water systems, which is where the greatest concentrations of chlorpyrifos were observed. This adds a degree of conservatism to the assessment.

Ecotoxicology is the study of toxicity-mediated effects on assemblages of two or more species of organisms (Cairns 1989). Thus, the use of distributions of species susceptibility is an extension of probabilistic approaches used for single species to assemblages of organisms. Just as the sustainability of a population of organisms may be maintained despite some effects on individual organisms in the population, the sustainability of population and community function can be maintained despite effects on some populations or communities. The probabilistic approach to ecotoxicological risk assessment simply allows the quantification of the likelihood of this occurring.

Assessment Criteria. When selecting assessment criteria for probabilistic risk assessments, several issues must be considered. In ecological risk assessment, it is recognized that individual organisms in the environment are part of the food web and are thus transitory, either consuming or being consumed (Suter et al. 1993). Thus, in general, ecological risk assessment is aimed at protection of the sustainability of populations and the functions of communities and ecosystems, rather than individual organisms. This acknowledges the fact that populations are less sensitive than their most sensitive member and, likewise, that communities and ecosystems are less sensitive than their most sensitive components. Thus, these criteria are better for assessment purposes. In this risk assessment of chlorpyrifos, the 10th centile of the sensitivity distribution was used as an initial working criterion. This criterion was critically evaluated against other approaches to risk assessment, such as FAVs and FCVs (USEPA 1995), a knowledge of the ecological role of the species most likely to be affected, and results from mesocosm and microcosm studies. Thus, the reference dose used in the probabilistic comparison was a consensus reference dose.

There are two basic methods of describing the relationship between exposure and response. One method is the hazard ratio technique used in the lower tiers of the risk assessment process. In this technique a single point estimate of both exposure and response is derived. This tends to bias the estimates such that hazards are either under- or overstated. To avoid underprotection, hazard quotient methods generally make very conservative estimates of both the exposure and the hazard values. In these initial tiers, the risk assessor is using a one-tailed test to try to screen out situations that are unlikely to result in adverse outcomes at the ecosystem or population level. Thus, the earlier tiers are designed to be protective, rather than predictive. Often, the hazard side of the relationship uses the most sensitive species, then adds additional safety factors to assure the risk assessor that there is little probability of finding a more sensitive species or endpoint. For this reason, the values selected as the point estimators have little or no probability of actually occurring in nature. While these methods can be used to determine the potential for hazard, because they do not consider the probability of exposure or response, they cannot be used to estimate risk, which always includes an element of probability. A practical consequence of these theoretical considerations is that it is difficult to estimate the concentrations that correspond to exceedences of effect concentrations for small proportions of populations.

The alternative to the hazard quotient methodology that is suggested for higher tiers of assessment is the probabilistic approach where all the information available to the risk assessor is used to determine the probability of overlap between the distributions of exposure and response. In this way, the assessor can provide the risk manager with probabilities of protecting a particular proportion of individuals or a particular proportion of species from a defined probability of occurrence of a particular magnitude of exposure. The intent of the probabilistic risk assessment is not to predict a "safe level," but rather to provide options for prioritization, mitigation, remediation, and restoration to the risk manager. While still estimates, this level of refinement allows the risk manager to put the potential for adverse effects into ecological perspective and apply risk–benefit decision criteria. The continuum of possible responses and exposures can be represented as a cumulative frequency distribution for which there are no zero or 100% values. These points are approached asymptotically, but can theoretically never be attained. The nature of the frequency distributions is such that it is meaningless to determine the concentration of a toxicant that would cause no effect. Instead, it is relevant to discuss the probability of exceeding the concentration required to affect a particular proportion of the individuals in a population of a defined proportion of species in a community. Another reason to place less reliance on the extreme tails of distributions is that, practically, there are thresholds of effects below which there are no measurable responses to toxicants, even for infinitely long exposures. The probability distributions are such that the rate of change in concentration to cause a particular response increases asymptotically as the cumulative frequency of including a species increases. For this reason, the change between the 98th and 99th centile is greater than the change between the 90th and 91st centile (see Fig. 27). The implication of this is that there is a point of diminishing returns. Based on the probability distributions, exposure concentrations would need to be infinitely small to include 100% of the possible individuals in a population or species in a community. When the exceedence profile is examined (Fig. 27), it can be seen that there is an apparent breakpoint, where the rate of change in concentration becomes relatively great compared to the rate of including additional more sensitive species. The 10th centile lies near the inflection point for many such distributions and can be defined with an acceptable level of discrimination, accuracy, and precision. Therefore, in an ecotoxicological risk assessment, the 10th centile can serve as a convenient reference point.

There are several other reasons why the use of the 10th centile of species affected in single-species, acute-toxicity tests can serve as a useful reference point for assessing the probability of response to a particular exposure. It is known from comparing the results of single-species tests to those of mesocosm and field studies that the concentration required to cause similar effects under simulated field conditions (in mesocosms) is greater (Graney et al. 1994), primarily because there is usually some moderation of the exposure under more realistic field conditions than in the laboratory. Thus, use of the 10th centile of the frequency distribution of effects as a reference point for risk assessment does not suggest that this would result in 10% of the species in the community being extirpated from an ecosystem by exposure to a toxicant present at this same concentration. Rather, the 10th centile is used as a reference exposure that can be calibrated with other outcomes and provide useful insight into the risk assessment process by identifying the more sensitive types of organisms in a community. Furthermore, use of this reference point does not suggest that the risk assessor believes that it would be acceptable to permanently lose 10% of the species from any community. The use of the probabilistic approach provides the risk assessor and risk manager flexibility in linking assessment and management. The probability of affected species may vary among ecosystems and jurisdictions. By using the process applied here, not only are the criteria and assessment endpoints more transparent, the endpoints can be changed without completely changing the assessment. If a greater level of assurance of protection of an ecosystem is desired, the centile chosen as the assessment endpoint as well as the measurement endpoint can be changed without the need to completely redo the exposure assessment. Furthermore, various endpoints can be chosen and the implications investigated before final decisions or actions are taken.

Ecosystems can be viewed from a structural or a functional perspective. Both have utility in the understanding of ecosystems. It can be held that any change in the absolute or relative numbers of individuals in a population or taxa in a community will result in a change in the ecosystem. However, practically, there are changes that can be accommodated without altering the functioning of the community that supports the structure of economically and aesthetically important higher trophic levels. These changes can be permanent or transient. Depending on the type of ecosystem, it may be acceptable to lose a portion of the individuals from a population transiently. For instance, organisms with short half-lives, rapid rates of intrinsic population increase, and resting stages or propagules are resilient and recover from short-term insults very quickly. During short periods of time when toxicants may have affected some sensitive individuals or populations, there is often functional replacement that allows ecosystems to continue functioning until the affected populations recover from short-term perturbations. For instance, for a substance that causes toxicity to some zooplankton, there may be increases in the populations of more tolerant organisms that can serve as a food supply for fish and other long-lived predators (Kaushik et al. 1986; Stephenson et al. 1986). In the case of chlorpyrifos, the longer-lived, k-selected species such as fish are less sensitive than the r-selected invertebrates. For this reason, if there were not functional substitution, annual short-term toxicity could extripate the longer-lived species as a result of the indirect effect of starvation. Because there are a sufficient number of tolerant replacement prey to bridge periods of depression of the more sensitive invertebrates, it is unlikely that this will occur in most surface water ecosystems. Because phytoplankton are not sensitive to exposure to chlorpyrifos, effects on primary production would not be expected. Release from grazing would also be unlikely to occur as more tolerant species of zooplankton could fill the role of these herbivores secondary producers over short periods of time.

In developing a reference concentration to compare to the exposure profiles in this assessment of chlorpyrifos, more information than just the 10th centile effect level was used. In addition, FAVs and FCVs were calculated using accepted USEPA methodologies. The EPA method used is a simplification of the probabilistic approach in which organisms are classified based on family and genus, instead of species. For this reason, the resolution of the EPA method would be expected to be less than the species-based probabilistic method. The exposure probabilities were compared to the results of three different methods of calculating reference doses (see Tables 21, 22).

The duration of the exposures in most streams is such that the acute lethality is the most appropriate value to which exposure concentrations should be compared. The reference concentration for chlorpyrifos based on the 10th centile of freshwater arthropod species was found to be 55 ng/L; the FAV, based on the USEPA methodology, was found to be 148 ng/L. Thus, the reference concentration predicted from the 10th centile of LC_{50} values, based on single-species toxicity tests, was approximately 3 fold more protective than would have been derived using USEPA methodologies. In addition, the MATC based on the 10th centile value was only approximately 2.5 times greater than the FCV, also derived by the USEPA methodology. Thus, if the exposure scenarios were more chronic in nature (>48 hr), it could be concluded that the 10th centile might be slightly underprotective. However, this was taken into consideration by examining the results of mesocosm studies that allowed observation of the response of communities and ecosystems under realistic field conditions. The results of mesocosm studies indicate that no ecologically relevant adverse effects are expected to occur at concentrations <100 ng chlorpyrifos /L. The reference concentration based on the 10th centile assumes 100% bioavailability of the compound and that all individuals are maximally exposed for the entire duration of the study. When all the evidence is taken into consideration, concentration of chlorpyrifos corresponding to the 10th centile of affected species provides a reasonable estimate of the reference concentration to protect populations of organisms from acute exposures.

If chronic chlorpyrifos exposures would have been expected, the same type of analysis could be used, but the endpoint selected for consideration might be an effect other than lethality, such as reproductive output. Also, the studies to be included in the development of the probability function would need to be of appropriate duration. Alternatively, an ACR could be applied to the reference concentration based on the arthropod 10th centile. The ACR for chlorpyrifos was found to be approximately 8. If this is applied to the arthropod 10th centile based on acute lethality, FCV is 6.8 ng chlorpyrifos/L, which is approximately three-fold less than the value determined based on the USEPA GLI methodology (20 ng/L). Thus, the use of the arthropod 10th centile of the probability distributions is a reasonable reference concentration and is as or more protective than the water quality criteria derived by accepted USEPA methods.

Lines of Evidence. The results of ecological risk assessments need to be interpreted in the context of ecological relevance. While the fact that the probabilistic approach is a purely numerical methodology is an advantage from the point of view of the transparency of the procedure, it cannot, nor is it designed to, assess the ecological relevance of the exceedences that may be identified. For example, an assessment criterion of the 10th centile may include keystone organisms of value to ecosystem function. Effects on keystone species, or long-lived, k-selected species, would be expected to extend to other species that are dependent on them, for example, as a source of food or as a predator. For this reason, it is necessary to assess the role of the potentially affected species in terms of their function in the ecosystem and whether this can be taken over by other organisms. The probabilistic approach can be used to refine the assessment process by allowing a rational ranking of scenarios by risk (likelihood of exceeding assessment criteria) and by identifying species in the distributions for which functional redundancy may exist (less sensitive organisms that can also perform the same function as the affected organisms). Ecological relevance can most usefully be assessed from a basic knowledge of ecology and from tests, such as microcosms, where community productivity and function can be evaluated directly. For this reason, refinement of the effects characterization in a probabilistic risk assessment gives a greater reduction of uncertainty. A number of assessment criteria can be compared to exposure distributions for the purposes of probabilistic risk assessments. These range from the 10th centile from the species probability plot for acute or chronic exposures through the point estimate for the most sensitive species to the conservative FCV, which is also actually derived in a probabilistic manner.

Return Frequency. The temporal scale of a stressor is important in ecological risk assessment. The return frequency of an event (how often the event happens) is an important consideration in the choice of methods for probabilistic risk assessment and is related to the ecological cost of recovery from the event (Solomon 1996). In assessing exposure, the return frequency of exposures from which organisms are protected should be consistent with the resiliency of vulnerable populations. Resiliency is determined by life cycle characteristics and reproductive capacity of the potentially affected organisms and the ability of their populations (or their function in the ecosystem) to recover from the episode. The report of the Aquatic Risk Assessment and Mitigation Dialogue Group (SETAC 1994) recommended conservative approaches to ecological risk assessment, such as the use of low return frequencies, for example, one or fewer occurrences in 30 yr. This was to safeguard all organisms in situations where limited information was available on mode of action or sensitivity of species.-

Where better information is available, more appropriate return frequencies may be used. For example, more frequent adverse events may be tolerated where a stressor affects organisms with short life cycles and high rates of reproduction. In phytoplankton, return frequencies of days to weeks will allow for recovery, even from high-impact events, especially if there is no persistent residue, as in the case for chlorpyrifos and other nonpersistent pesticides. Similarly, zooplankton and aquatic plants also would be protected by somewhat longer return frequencies; however, these would still be less than a season. In temperate regions many ecosystems undergo a period of dormancy and the system is, in a sense, reset seasonally by the winter. Thus, for some organisms, mechanisms for propagation beyond the winter reset already exist, and resting and other dormant stages are produced from which the next season's populations will develop. Similar mechanisms exist in environments with a dry season where ephemeral water bodies are subjected to drying out. Therefore, as many organisms in these regions undergo seasonal resets anyway, a stressor return that occurs less frequently than once per season is likely to be tolerable from the viewpoint of the long-term productivity of the population and the sustainability of function in the ecosystem. Protection of longer-lived species without seasonal resets, such as some fish, birds, or mammals, may however require the consideration of return frequencies of several years.

Refugia. If a stressor is present nonuniformly in the environment, unexposed areas will act as refugia for repopulation of potentially impacted areas. The relative size of the exposed and unexposed areas is important, but this issue is particularly significant for assessing risks from pesticide use, where untreated fields, set-aside land, crop rotations, and mixed farming practices guarantee that refugia will be present. Similarly, refugia exist in streams and rivers, and many organisms have resistant stages or propagules from which population recovery can occur. Thus, probabilistic risk assessments (and hazard quotients) are additionally conservative because they do not consider repopulation from unexposed refugia. The example of the more rapid than expected recovery of the biota in the River Rhine from an endosulfan spill illustrates this point (Friege 1986).

B. Results of the Risk Analysis

Hazard Quotient Analyses. The hazard quotient method (HQ) of risk analysis (Urban and Cook 1986) compares an expected environmental concentration (EEC) to a measure of hazard to aquatic organisms (for acute effects, typically the LC_{50} for a sensitive species). Three hazard categories are defined for nonendangered species (Eq. 8):

$$HQ = EEC/LC50 \qquad (8)$$

For $HQ < 0.1$ Presumption of no hazard (10 × safety factor)
For $0.1 \leq HQ < 0.5$ Presumption of hazard that may be mitigated by restricted use (2 × safety factor)

For HQ ≥ 0.5   Presumption of unacceptable hazard

Interpretation of the hazard categories can be simplified by applying the 2 × safety factor to the hazard value, which results in a ratio that generates a presumption of acceptable hazard for all HQ ≤ 1 (Eq. 9):

$$HQ = EEC/(0.5 \times LC_{50}) \tag{9}$$

Any use pattern generating an EEC that results in HQ > 1 is determined to have exceeded a level of concern (LOC) at which hazard to aquatic organisms is assumed.

Tier I. A Tier I exposure assessment for a T-band application of 1 kg/ha (≈ 1.2 lb a.i./acre) in field corn was performed using the EPA GENEEC computer program and the chemicophysical properties of chlorpyrifos. Hazard quotients were calculated based on the *Gammarus pulex* LC_{50} because *Gammarus* was one of the aquatic genera most sensitive to chlorpyrifos and *G. pulex* was the most sensitive nontarget species. The use of acute toxicity information is justified because exposures are generally less than 48 hr and, if exposure to chlorpyrifos does not result in death, organisms recover completely. Using an acute LC_{50} value of 70 ng/L for *G. pulex* to represent toxicity to freshwater aquatic invertebrates and applying the 2-× uncertainty factor, the following hazard quotients were obtained. The HQ for peak concentrations was 73 (2560/70/2), while the 96-hr average HQ was 62 (2160/70/2). The assessment for freshwater fish, using an acute LC_{50} of 1300 ng/L for the most sensitive fish, carp (*Cyprinus carpio*), resulted in a peak HQ of 3.9 and a 96-hr average HQ of 3.3.

These results indicate that the T-band use of chlorpyrifos would result in levels of concern (LOCs) that would trigger a more refined risk assessment beyond the Tier I screen for invertebrates or fish, and the exposure assessment should be refined in Tier II. Similar results would be expected from other uses of chlorpyrifos. The Tier I assessment is a one-tailed screening-level analysis designed to eliminate the need for complex risk assessments for compounds that represent negligible ecological risk. This is not the case for chlorpyrifos. This analysis also indicates that fish are more tolerant of the effects of chlorpyrifos and that protecting sensitive invertebrates will also protect fish from the direct and indirect adverse effects of chlorpyrifos. The greatest uncertainty in the Tier I exposure assessment is caused by the conservative assumptions made by the GENEEC model. Reduction in the uncertainty in the exposure assessment can be accomplished by the use of more sophisticated exposure models as outlined in Tiers II and III or the use of actual final monitoring data (SETAC 1994).

The use of the most sensitive test result for the most sensitive species of nontarget organism results in a high degree of conservatism. If the foregoing hazard quotients were calculated with genus mean acute values (GMAVs) for tests selected according to the criteria for the calculation of FAVs (see Table 21), more appropriate HQs may be obtained. For example, if the LC_{50} for *Fundulus similis* (5789 ng/L) was used in the HQ calculation, an HQ of 0.88 for fish would have been obtained. If the GMAV for the next most sensitive freshwater fish were used, the HQ would be even less.

Tier II. This assessment introduced some aspects of probability into the analysis by using a PRZM/EXAMS simulation of runoff and selecting a 90th centile weather pattern to drive the runoff events and by considering two sites to represent typical and "worst-case" site vulnerabilities. For a banded application rate of 1.16 kg/ha ($\approx$ 1.3 lb a.i./acre), typical worst-case exposures and hazard quotients were obtained, again using the 96-hr *Gammarus pulex* LC_{50} (70 ng/L) and the carp 96-hr LC_{50} (1300 ng/L), with an uncertainty factor of 2.

The results suggest that most of the corn uses of chlorpyrifos (and others for which similar assessments were performed) would result in HQs that exceed arthropod and fish LOCs in the Tier II exposure assessment, and additional refinement is necessary (Table 25). The safety factor used in the assessment would result in an approximately twofold overestimation of risk. This, combined with the fact that fish are much less sensitive than invertebrates, suggests that there would be little risk of adverse effects from exposure of fish to EECs derived from either scenario. There would, however, be some LOC for invertebrates for both scenarios.

Tier III. Although the Tier III assessment suggested that the risk to aquatic organisms was less compared to the Tier II assessment, it still predicted an unacceptable LOC for invertebrates in all the Midwestern corn-growing area in conventional tillage (T-band application of 1.07 kg/ha [$\approx$ 1.2 lb a.i./acre]). Hazard to fish was estimated to be unacceptable on about 20% of the area (Table 26). However, few fish kills have been reported that were specifically related to use of chlorpyrifos in corn (USEPA 1996b). This suggests that the edge-of-field, regional-scale runoff modeling in the Tier III exposure assessment may overestimate EECs or that there are other factors mitigating the toxicity of the predicted EECs. Factors that may contribute to overestimation of the EEC include conservative, typical worst-case assumptions for weather, field drainage, and possible mitigation by buffer zones at field borders. There is also minimal dilution and dispersion of pulsed inputs into the constant-volume, static farm pond, which is not representative of the large reservoirs or flowing water sys-

Table 25. Tier II EECs and hazard quotients (HQ).

Site	Time	*Gammarus pulex*		*Cyprinus carpio*	
		EEC (ng/L)	HQ	EEC (ng/L)	HQ
Iowa	Peak	2,800	80	2,800	4
	96-hr	2,300	66	2,300	4
Mississippi	Peak	13,000	371	13,000	20
	96-hr	10,200	291	10,200	16

Table 26. Tier III EECs and HQs.

Percent of simulated area[a]	Cumulative percent of simulated area	Invertebrates		Fish	
		EEC (ng/L)	HQ	EEC (ng/L)	HQ
25.2	25.2	100	3	100	0.2
21.2	46.4	200	6	200	0.3
14.9	61.3	300	9	300	0.5
8.4	69.7	400	11	400	0.6
6.8	76.5	500	14	500	0.8
5	81.5	600	17	600	0.9
3.6	85.1	700	20	700	1.1
3.2	88.3	800	23	800	1.2
2.1	90.4	900	26	900	1.4

[a]Total number of acres simulated = 2 million ha (249,740,324 acres).

tems that may be considered as socially more important ecosystems requiring greater protection. The capability of existing environmental fate models to give accurate edge-of-field predictions is also in question (Solomon et al. 1996). While use of the Tier III assessment would be protective, it is questionable that it is predictive of actual field exposures. Additional refinement of the exposure assessment is necessary to reconcile LOCs of numerical risk assessments with field observations that indicate that predictions of adverse effects are overestimated. Edge-of-field, regional-scale runoff modeling in the Tier III exposure assessment may overestimate EECs or that there are other factors mitigating the toxicity of the predicted EECs. Additional results suggesting EEC overprediction were obtained in similar Tier III assessments conducted by Dow-Elanco for chlorpyrifos use in cotton, peanuts, sugar beets, and tobacco (unpublished internal DowElanco reports).

Probabilistic Risk Analyses of Measured Concentrations in Freshwater. Concentrations of chlorpyrifos in surface freshwater available from several monitoring programs were used in a region-specific probabilistic risk assessment. Four criteria were used to assess the likelihood of exposure exceeding a reference assessment measure. These were the 10th centile of 48-hr LC_{50}s for all organisms for which acute toxicity information was available (102 ng/L); the freshwater FAV (148 ng/L); the final chronic value (FCV) (18 ng/L) derived from an ACR based on both freshwater and saltwater organisms; and the freshwater mesocosm NOAEC (100 ng/L). The FCV is a conservative endpoint because it is protective of chronic exposures that would be unlikely to occur in the case of chlorpyrifos. It is also less than the 10th centile for FW arthropods (55 ng/L), which were the most sensitive organisms. The 10th centile of the acute sensitivity distribution of organisms has been used in previous risk assessments (Solomon et al. 1996; Solomon and Chappel 1997) and is consistent with the NOAEC concentration observed in mesocosm studies with chlorpyrifos. Compared with the 10th centiles of the toxicity distributions for fish, these criteria are conservative and will, by default, also protect fish from direct toxicity.

The likelihood that either the 10th centile of the acute toxicity distribution or the NOAEC from the mesocosms would be exceeded in any of the Lake Erie watersheds (Table 27) was small, less than 10% in all cases. Only in the Huron River was the FCV exceeded for more than 10% of the sampling intervals. This exceedence is unlikely to be of sufficient duration to cause chronic responses. Thus, the assessment would be conservative.

Exceedences in the California sampling data (Table 28) demonstrated two trends. The concentrations of chlorpyrifos in the mainstem rivers have a small probability of exceeding any assessment criteria. However, in the smaller tributaries and agricultural drains, there was a greater likelihood that all the criteria would be exceeded. Similar results were observed in the case of atrazine (Solomon et al. 1996) where concentrations in small streams and reservoirs draining intensively managed farmland were more likely to exceed assessment criteria than were concentrations in large rivers. In situations such as those in the smaller streams in agricultural regions of California, the results of the refined risk assessment (Tier IV) indicate that adverse effects of concentrations of chlorpyrifos in these systems is possible. Therefore, site-specific risk assessments should be carried out in the context of the designated uses of these systems. The probabilities of exceeding the assessment measure were very small. The results of mesocosm studies where transient effects on invertebrates were observed did not result in direct or indirect effects on fish populations observed at the same exposures. Thus, even in the California streams fish would not likely be adversely affected at most concentrations observed in agricultural regions. This conclusion is consistent with the fact that few fish kills in the regions studied can be attributed to the use of chlorpyrifos following label instructions. While lethality of fishes would not be expected, even at the greatest concentrations of chlorpyrifos observed, toxicity to primary consumers could cause indirect effects on fish, especially on life stages sensitive to nutritional limitations, particularly young fish.

Assessment of the USGS National Water Quality Assessment Program (NAWQA) data set (Table 29) revealed trends similar to the California data with respect to watershed size. The probability of the 10th centiles for the acute responses or the mesocosm NOAECs being exceeded was less than 10% in all cases, except for the Cherry Creek, near Denver, CO. Probabilities of exceeding the FCV were greater than 10% in several other locations (Table 29), but these exceedences were not restricted to either agricultural or urban indicator sites only. Greater probabilities of exceedences in the Nebraska (Shell Creek and Prairie Creek) and California (Orestimba Creek and the Merced River) are consistent with the greater use of chlorpyrifos in these areas. The months during

Table 27. Probability of concentrations of chlorpyrifos in the Lake Erie drainage watersheds exceeding various assessment measures.

Watershed	Years monitored	Mean no. daily observations/year	Percent probability of exceeding:			
			FW acute all organism 10th centile (102 ng/L)	FW FAV (148 ng/L)	FCV (18 ng/L)	FW Mesocosm NOEC (100 ng/L)
Lost Creek, OH	1983–1990, 1992–1993	52	0.5	0.2	4	0.5
Rock Creek, OH	1983–1995	64	0.9	0.3	7	1.0
Honey Creek, OH	1983–1995	70	0.5	0.1	5	0.6
Huron River, OH	1988–1991	48	5	4	14	5
River Raisin, MI	1983–1995	64	2	2	7	2
Sandusky River, OH	1983–1995	67	5	5	10	5
Maumee River, OH	1983–1995	64	2	2	7	2

FW, freshwater; FAV, final acute value; FCV, final chronic value; NOEC, no-observed-effect concentration.

Table 28. Probability of concentrations of chlorpyrifos in the California watersheds exceeding various assessment measures.

		Probability of exceeding:			
Location	N	FW acute all organism 10th centile (102 ng/L)	FW FAV (148 ng/L)	FW flow-through (18 ng/L)	Mesocosm NOEC (100 ng/L)
Salt Slough	10	10	9	16	10
Orestimba Creek	19	23	17	54	23
Spanish Grant Combined Drain	19	31	23	70	31
T.I.D. No. 3	17	33	27	62	33
T.I.D. No. 5	22	15	10	54	15
T.I.D. No. 6	17	24	18	55	24
Del Puerto Creek	22	3	1	38	3
Ingham/Hospital Creeks	24	13	10	38	13
Merced River	17	10	7	38	11
Tuolumne River	11	0.3	<0.1	21	0.3
San Joaquin River at Hills Ferry Road	37	0.2	<0.1	5	0.2
San Joaquin River at Laird Park	122	1.8	1	9	1.9
San Joaquin River at Airport Road	14	0.8	0.3	21	0.8
San Joaquin River at Vernalis	1127	<0.1	<0.1	0.5	<0.1
Alamo River at Harris St. Bridge	15	19	19	45	20

Table 29. Probability of concentrations of chlorpyrifos in the NAWQA data sets exceeding various assessment measures.

			Probability of exceeding the following assessment measures:			
Location	Type	N	FW acute all organism 10th centile (102 ng/L)	FW FAV (148 ng/L)	FCV (18 ng/L)	FW Mesocosm NOEC (100 ng/L)
splt-cherry	indicator-urban	36	9.6	7.4	26.4	9.7
cnbr-prairie	indicator-agric.	16	8.8	7.5	17.4	8.9
sanj-orest	indicator-agric.	85	7.7	4.6	42.3	7.9
cnbr-shell	indicator-agric.	15	7.4	5.5	23	7.5
trin-rush	indicator-urban	22	6.5	3.3	49.9	6.7
sanj-merced	indicator-agric.	40	4.8	3.7	14.1	4.9
nvbr-lasvegas	indicator-urban	50	5	4	54	5
splt-lone	indicator-agric.	33	3.9	2.6	18.5	4.0
ccpt-crab.rl	indicator-agric.	35	4	4	10	4
will-zollner	indicator-agric.	29	3	3	31	3
ccpt-el68	indicator-agric.	31	3.0	1.8	19.1	3.1
cnbr-platte	Integrator	59	2.6	1.8	10.6	2.6
cnbr-maple	indicator-agric.	59	2	2	21	2
poto-mono	indicator-agric.	45	1.5	0.3	14.1	1.6
whit-kess	indicator-agric.	36	1.6	1.2	5.5	1.6
whit-white	Integrator	72	0.8	0.5	6.0	0.8
lsus-eastm	indicator-agric.	96	0.6	0.2	18.8	0.6
acfb-sope	indicator-urban	52	0.6	0.2	14.5	0.6

Site	Type	n			
whit-little	indicator-urban	44	0.6	6.6	0.6
poto-accotink	indicator-urban	28	0.4	6.9	0.5
whit-sugar	indicator-agric.	37	0.4	5.8	0.5
will-pudding	indicator-agric.	72	0.4	3.7	0.4
All Integrator	Integrator	30	0.4	10	0.4
lsus-mill	indicator-agric.	43	0.2	4.0	0.2
redn-snake	indicator-agric.	27	<0.1	7.0	<0.1
sanj-salt	indicator-agric.	26	<0.1	30.4	<0.1
will-famo	indicator-urban	28	<0.1	32.8	<0.1
usnk-rock	indicator-agric.	41	<0.1	1.0	<0.1
albe-pete	indicator-agric.	35	<0.1	0.4	<0.1
gafl-tucsa	indicator-agric.	53	<0.1	<0.1	<0.1
sanj-sj	Integrator	73	<0.1	18.3	<0.1
gafl-lafayette	indicator-urban	41	<0.1	5.1	<0.1
gafl-withla	Integrator	37	<0.1	<0.1	<0.1
lsus-cedar	indicator-urban	61	<0.1	3.9	<0.1
All indicator-agric.	indicator-agric.	479	1.4	13.8	1.4
All indicator-urban	indicator-urban	1342	1.4	8.3	1.4
All sites	—	2194	1.2	8.9	1.2

which greater concentrations occurred in these locations were consistent with use in corn in Nebraska (June) and with use of chlorpyrifos in tree crops and alfalfa in California (March through May).

The exceedences of the annual maximum measured concentrations (see Table 12) were compared to the 10th centile of the distribution for freshwater (FW) fish (Table 30). Ideally, a distributional approach would have been used; however, the data sets were judged to be too small for this purpose and a simple quotient ratio was therefore used. In no case was the 10th centile of the fish toxicity distribution exceeded, confirming the judgment that, even over a period of several years, chlorpyrifos concentrations in this drainage basin would be unlikely to exceed concentrations at which fish might be affected directly. By extension, amphibians were also judged to be unlikely to be affected; however, this conclusion has greater uncertainty because of a lack of appropriate effects data for this class of organisms. When these annual maxima were compared to the 10th centiles of effects on FW arthropods, they suggest that, in some rivers in this drainage basin and in some years, significant effects on invertebrates would be predicted to occur. These effects must be considered in the context of the ability of these groups to sustain higher return frequencies than longer-lived organisms such as the fish discussed earlier and other factors that may be limiting populations of invertebrates in streams of these agricultural areas.

Probabilistic Risk Analyses of Measured Concentrations in Saltwater. Concentrations of chlorpyrifos measured in Chesapeake Bay (see Fig. 18) were small and, in all cases, the likelihood of exceedence of any of the assessment measures was <0.1%. Based on this single estuary, which receives water from a number of agricultural areas in which chlorpyrifos is used, suggests that the risk to saltwater organisms in estuarine environments is small. However, the little information available for estuaries makes this conclusion uncertain.

Table 30. Hazard quotient for the yearly annual maximum measured concentrations of chlorpyrifos in the Lake Erie drainage basin based on the 10th centile for FW fish (5358 ng/L).

Year	Maumee	Sandusky	Raisin	Huron	Honey	Rock	Lost	Cuya-hoga
1987	0.02	—	0.02	—	0.03	—	0.03	0.01
1988	0.05	—	0.05	0.1	0.01	0.07	0.02	0.09
1989	0.01	0.04	—	0.18	0.02	0.01	0.02	0.01
1990	0.01	0.46	—	0.09	0.01	0.06	0.03	0.01
1991	0.09	0.29	—	0.07	0.01	0.15	0.00	0.01
1992	0.21	0.24	0.01	—	0.06	0.01	0.10	—
1993	0.04	0.53	0.00	—	0.04	0.04	0.12	0.01
1994	0.04	0.21	—	—	0.04	0.03	—	—
1995	0.03	0.06	0.02	—	0.04	0.04	—	0.02

Exceedence Profile Plots. Exceedence profiles were constructed for selected sites by combining the frequency distributions for the probability of a concentration being exceeded with the probability of a species being affected to generate a graphical expression of the percent exceedence of LC_{50}/EC_{50} values against the probability of a species being affected (Fig. 28). The TID-3 Turlock Irrigation Ditch site in California was chosen as a worst case site representative of the small creeks and drains in this use area. The San Joaquin River was chosen to represent larger river systems and the Huron River was chosen as a worst case site typical of the corn use area. An additional advantage of the probabilistic methods applied in the ecological risk assessment of chlorpyrifos is that the analysis is independent of the *a priori* selection of acceptable probabilities of exposure and response for the decision-making process. The values chosen for the exceedence probabilities of exposure and effect can be selected on the basis of the degree of protection required for particular situations. The concentration of chlorpyrifos associated with the 10th centile of species sensitivities was determined to be 102 ng/L, which is essentially the same as the NAOEL derived from mesocosm studies. For this reason, the use of the 10th centile is suggested as a useful assessment endpoint, especially when information from mesocosm studies is not available. By examining the exceedence profiles (Fig. 28), it can be seen that, when considering all species, the reference concentration based on the 10th centile would not be exceeded at the Laird Park, CA, sampling station on the San Joaquin River, or at the Huron River site in Ohio, but would be exceeded at the TID-3 location on the San Joaquin system. If only fish are considered, the MATC based on the 10th centile would not be exceeded at either the Laird Park or TID 3 locations in the San Joaquin.

If other levels of effect, expressed as the probability of species affected (lesser centiles) or the FCV had been selected as the reference point, neither the ranking of relative risks nor the conclusions drawn would change. As the refer-

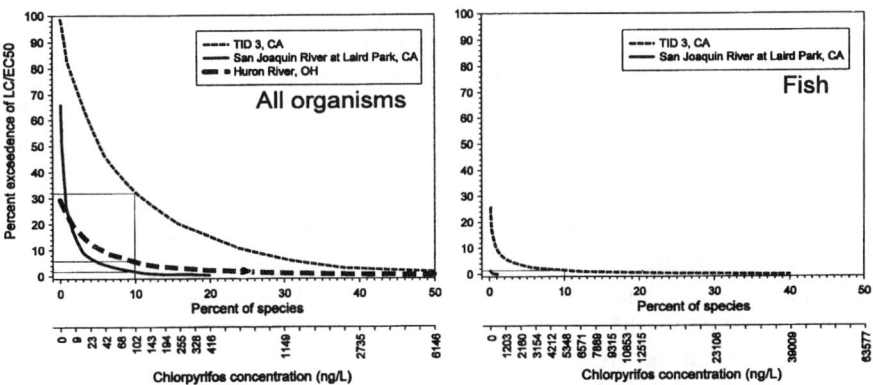

Fig. 28. Chlorpyrifos exceedence profiles for species distributions of all organisms at selected sites.

ence concentration based on the 10th centile was used as a reference point, it is instructive to compare the conclusions that would be drawn based on reference concentrations based on other degrees of response. In the assessment of chlorpyrifos, this was done by comparing the reference doses based on the 10th centile as well as the FAV and FCV and the microcosm NOAEC (see Tables 27–29). Similarly, if different proportions of potentially affected species were selected, the conclusions would not be substantially different. For instance, the reference concentrations based on the 5th and 1st centiles of all species for which single-species toxicity testing information was available were found to be 35 and 3.58 ng/L chlorpyrifos, respectively. While the lesser probabilities of exceedence are more protective, estimates of reference concentrations based on these endpoints are also more uncertain and often beyond the level to which extrapolations should be made. For instance, if there are only 20 species included in the data set used to develop the exceedence relationships, extrapolations to less than the 5th centile should not be made without due consideration of the uncertainties that would result.

The risk assessment for chlorpyrifos was not designed to prove that there was no risk to aquatic organisms, or even to predict the level of risk, but rather as a guide in decision making by allowing prioritization of the potential risks at different locations and to eliminate from further consideration species and locations that were very unlikely to be at risk. All the reference concentrations were used to identify ecosystems potentially at risk and prioritize those possibly deserving additional study. These studies could include collection of additional information to refine the risk assessment, site-specific monitoring, or potentially mitigation measures. The refined risk assessment presented here does not demonstrate that chlorpyrifos is causing adverse effects on communities, but rather allows the prioritization of locations where additional information may be needed to make a decision about the safety of the use of chlorpyrifos in these regions. The use of site-specific studies of community structure and function may be appropriate in those areas where the probabilistic assessment indicated the risks to be greatest. However, the use of field studies to determine the potential for effects of individual agricultural chemicals on aquatic communities is complicated by the fact that organisms are exposed simultaneously to a number of chemical and nonchemical alterations of these environments. It would be difficult, without application of the appropriate experimental design, to separate the effects of one chemical or nonchemical alteration from another. Streams, which were the primary focus of this risk assessment for chlorpyrifos, that occur in agricultural regions have been greatly altered from their primordial state. Typically, the streams are more turbid and warmer than they would have been before alteration of the landscape by agriculture. An appropriate strategy for assessing the potential for adverse effects would be to investigate the systems predicted to be at the greatest risk. If conditions at these locations were found to be protective of the species present, by logic, locations with lesser probabilities of overlap would be expected to be protected.

Use for Termite Control. Although environmental concentrations of chlorpyrifos related to its use in termite control are not available, some survey data were available for this use pattern. Between 1993 and June 1996, 44 suspected incidents involving surface water contamination from termiticide use across the U.S. were confirmed by chemical analysis, investigated, and documented in an internal survey conducted by DowElanco (Thomas and Chambers 1996). This type of monitoring is nonrandom in nature because of reliance on suspicion of difficulties with the application (odor, observation of emulsion, fish kill, etc.). Despite this limitation, incident reports are quite valuable in assessing the frequency of occurrence of contamination for this use pattern, because all treatments are professionally applied by businesses that are encouraged to remediate problems when they develop and to report these incidents to the manufacturer. However, minor incidents that do not generate suspicion of problems will not be captured by this type of incident monitoring, and it is possible that contamination at concentrations toxic to sensitive aquatic species may occur without being reported. Incidents of contamination of surface water from termite control applications (Thomas and Chambers 1996) (Table 31) showed that states which experienced the greatest numbers of treatments also reported the fewest annual

Table 31. Termiticide incidents for the period January 1993 through June 1996.

State	Percent of total treatments	Annual incident rate (%)
WA	0.08	0.0138
IA	0.52	0.0041
MI	0.88	0.0025
PA	1.49	0.0088
AR	2.02	0.0016
IL	2.63	0.0058
KY	2.65	0.0016
MS	2.83	0.0011
CA	2.87	0.0004
LA	3.06	0.0007
IN	3.09	0.0070
MO	3.32	0.0013
SC	3.63	0.0012
OH	4.11	0.0013
TN	4.27	0.0008
NC	5.88	0.0011
GA	7.03	0.0014
TX	8.01	0.0005
AL	8.13	0.0008
FL	10.15	0.0007

incident rates. The range of rates was from 0.00034% to 0.0138%, or 0.34–13.8 per 100,000 applications.

Although monitoring data were not available for these sites, the probability of an incident involving reports of fish kills or noticeable contamination of surface water appears to be small overall and especially small in those states with maximum use against termites. Thus, in some cases, chlorpyrifos concentrations were great enough to cause fish kills but the overall probability of these occurring on the scale of the landscape and use pattern was small. In addition, many of these incidents resulted in localized contaminations on the property where the termite control operation was being conducted and therefore need to be assessed in the context of the risks and benefits to the property owner.

VI. Ecological Significance of Effects

Interpretation of the foregoing risk analysis for chlorpyrifos must consider a number of issues relating to the significance of the results in the context of ecosystem relevance. As recommended in the Framework for Risk Assessment (USEPA 1992a, 1996a), risk assessment must consider all the lines of evidence, and this information needs to be weighed in the context of an understanding of ecological processes and their relevance in the ecosystem. These are discussed in the following sections.

A. Effects Criteria

The effects data upon which this assessment was based consisted mainly of LC_{50} measurements, primarily because these are the most commonly available data. It has been suggested that lesser (benchmark) effect levels such as the LC_{10} or the LC_5 may be more appropriate for these types of risk assessments (SETAC 1994). However, this recommendation was designed to consider some groups of compounds for which the slope of the concentration–response relationship is small. Typically, these relationships have been noted for toxicity studies where organisms not possessing the receptor system for the pesticide have been tested, for example, the effects of fungicides on crustacea. Most (95%) concentration–response slopes for insecticides tested in fish are greater than 1.4, and the median slope has been reported to be 5.0 (SETAC 1994). Under these conditions, the differences between an LC_{50} and a smaller effect level, such as LC_5, would be small. In these circumstances, the use of the LC_{50} can be justified, particularly if the exposures are of short duration and there is little cumulative adverse effect, as is the case for chlorpyrifos. Also, the results of the assessment were supported by results from mesocosm studies, which represent longer-term and more realistic exposures of communities.

Studies of effects of insecticides on zooplankton and miticides in terrestrial arthropods have shown that LC_{50} values are not a good predictor of effects at the population level (Daniels and Allen 1981; Day and Kaushik 1987; Walthall and Stark 1997). Day and Kaushik (1987) showed that *Daphnia galeata men-*

dota populations exposed to sublethal concentrations of the pyrethroid insecticide fenvalerate were able to sustain a rate of increase similar to that of unexposed controls. Similar results have been observed for the insecticide dieldrin in *Daphnia pulex* (Daniels and Allen 1981). Working on *Acyrthosiphon pisum* and its response to the miticide imidacloprid, Walthall and Stark (1997) showed that the populations exposed to the 72-hr LC_{60} were able to maintain rates of population increase similar to untreated controls. This lack of population-level responses, even at exposure concentrations greater than the LC_{50}, has been attributed to compensatory mechanisms by which the unaffected individuals are able to maintain heightened rates of reproduction by decreased competition for limiting resources (Walthall and Stark 1997). Results of this type support the conclusion that use of the results of single-species acute toxicity studies with lethality as the endpoint results in conservative effects endpoints.

From a theoretical point of view, any effects criterion could be used for assessment purposes, provided that this criterion can be validated against knowledge and understanding of ecosystem structure and function. The effects of chlorpyrifos on aquatic organisms have been assessed in a number of mesocosm tests. These results demonstrated a community-level NOAEC of 100 ng/L, which is similar to the 10th centile intercept of the distribution of LC_{50}s of all organisms and greater than that from arthropods from laboratory studies. This conservatism of the 10th centile of LC_{50} data has been observed in other situations (Solomon et al. 1996) and, among other reasons, is probably a result of the population-level mechanisms discussed earlier and the maximization of exposure potential in laboratory studies conducted in sediment-free waters under flow-through or static conditions. While the results of the mesocosm studies should take precedence over those of the single species, in laboratory acute lethality tests, when they are available, the results with chlorpyrifos indicate that use of the 10th centile for other compounds for which extensive mesocosm studies have not been conducted is justified.

B. Ecological Role of Sensitive Taxa

The probabilistic ecological risk assessment of chlorpyrifos demonstrated that, at certain times in certain water bodies, concentrations of chlorpyrifos in the water were likely to exceed the 10th centile of all species sensitivities (102 ng/L) or the FAV (148 ng/L). To better understand the possible ecological implications of such exceedences, the tolerances of individual species were examined to determine which taxa were at greatest risk. This analysis focused on invertebrates, which are more sensitive to chlorpyrifos than are fish. For each of the major groups of freshwater invertebrates, the geometric mean of the 48-hr LC_{50} values was calculated. The 16 most sensitive freshwater invertebrates, which had 48-hr LC_{50} values of 300 ng/L or less, were identified (Table 32). The 16 most sensitive invertebrate species include 4 mosquito species, 4 cladocerans, 5 amphipods, 2 midges, and an odonate (dragonfly). Mayflies, caddisflies, hemiptera (backswimmers and water boatmen), beetles, diptera other than mosquitos

Table 32. Toxicity of chlorpyrifos to freshwater invertebrate groups

Taxonomic group	Geometric mean 48-hr LC_{50} (ng/L)	N	Number of species with 48-h $LC_{50}s \leq 300$ ng/L
Mosquitoes	100	8	4
Amphipods	370	8	5
Cladocerans	541	9	4
Coleoptera	1,125	3	0
Odonates	1,327	3	1
Mayflies	1,882	4	0
Midges	2,555	7	2
Caddisflies	3,884	2	0
Stoneflies	5,013	4	0
Other Diptera	5,692	4	0
Isopods	10,392	2	0
Hemiptera	10,455	4	0
Crayfish	18,814	2	0
Ostracods	23,270	3	0
Other invertebrates	48,165	3	0
Molluscs	654,968	7	0
Rotifers	8,449,928	1	0

and midges, stoneflies, mollusks, and rotifers are all more tolerant, with geometric mean 48-hr $LC_{50}s$ for each taxon ranging from 1,126 to 8×10^6 ng/L. The other two odonates in the database also have 48-hr $LC_{50}s$ in this range (4,101 and 8,061 ng/L). Crustacea other than cladocera and amphipods (isopods, ostracods, and crayfish) are among the most tolerant invertebrate groups, with geometric mean $LC_{50}s$ greater than 10,000 ng/L. Snails and rotifers, with geometric mean $LC_{50}s$ greater than 500,000 ng/L, are, as a group, tolerant to the effects of chlorpyrifos. These observations imply that chlorpyrifos exposure events in the streams and rivers for which monitoring data are available are unlikely to affect taxa other than mosquitos, midges, cladocerans, and amphipods.

Streams are not typical habitats for some of the sensitive TAXA such as mosquito larvae. Cladocerans also typically inhabit standing water ecosystems, and are less important in the ecology of flowing waters. Ecological effects of chlorpyrifos exposures at the upper end of the exposure distributions represented by the monitoring data are therefore likely to be limited to amphipods and midges. Little or no impact on mayflies, caddisflies, and stoneflies, which often constitute the majority of taxa that occur at the greatest density, is likely. Reductions in amphipod and midge populations, if they occur, are not likely to result in significant changes in the overall function of a stream ecosystem. These organisms are typically herbivores and detritivores, and their ecological functions are duplicated by many of the other invertebrate groups that would not be affected by chlorpyrifos (such as isopods, ostracods, and snails). Amphipods and midge larvae are part of the food supply for some fish and other predators, but a wide range of alternative food sources would remain even if these more sensitive taxa were reduced. In small streams, in fact, the most important component of the diet of fish may be terrestrial organisms that fall into the water. Furthermore, these organisms have rapid life cycles, often as short as 21 d, and relatively great fecundities. For this reason, short-term depressions of populations of these species are unlikely to result in permanent alterations in community structure or function. So long as there are sufficient numbers of prey items available to sustain predators to bridge the depression of some primary consumers, it is unlikely that higher trophic level organisms would be adversely affected by indirect effects caused by short-term exposures to chlorpyrifos.

C. Spatial and Temporal Issues

Distributional analyses of measured exposures can consider both spatial and temporal distributions of environmental concentrations. This allows exposure assessments to be refined to identify locations or times at which risks may be greater than at others. However, implicit in the concept of spatial and temporal distributions is the fact that the probability of co-occurrence of the sensitive organisms and the greater concentrations of the stressor may, in fact, be small. The distributional approach assumes that all the species may be exposed to the substance being assessed whereas the likelihood of this is small in a particular location. Similarly, it is well known that some organisms are more dominant at certain times of the season than at others or in some locations rather than others. Coincidence of dominance and greater exposure concentrations at a particular location could therefore increase risk in some situations but reduce it in others. This must also be viewed in the context of natural variability in spatial distributions of species in the landscape of the ecosystem.

Also relevant to this discussion is the issue of recovery from refugia. Because spatial distributions of chlorpyrifos and sensitive populations are not necessarily contiguous in the ecosystem, unexposed populations of organisms will exist. These populations act as sources from which repopulation can occur. An example of such a system would be organisms in a tributary of a river or stream from an unexposed region. Any habitat void caused in response to high concentrations of chlorpyrifos will be rapidly filled from these refugia. This event has been observed, even after such catastrophic events such as a pesticide spill in a river (Friege 1986).

D. Return Frequency and Recovery of Affected Populations

As discussed, return frequency of high-exposure, high-risk events must be assessed in terms of the species present in the ecosystem and their reproductive and stressor avoidance strategies. Fish (and amphibians) generally have a life cycle of 1 yr or longer. Thus, if they are subjected to even a single exposure of chlorpyrifos great enough to cause widespread mortality, the population may be adversely affected for up to several years. If such incidences occur every year or every other few years and there is no readily available colonization source, species of this type can be completely extripated from an ecosystem. While large enough exposure data sets were few in this assessment of chlorpyrifos, the data from the Lake Erie drainage basin spanned up to 13 yr of sampling. Chlorpyrifos use in the region increased during the period. Thus, the early years of data are not representative of current or future use practices, and distributional analysis of annual maximum concentrations were judged inappropriate, however, none of the measured annual maxima (see Table 12) exceeded the 10th centile of the distribution of the FW fish LC_{50}s. In fact, even the greatest annual maximum chlorpyrifos concentration reported was only approximately one-half the 10th centile of the fish LC_{50} distribution, and many arthropods also would have been unaffected at this concentration. The narrow range of the distribution of FW fish LC_{50}s suggests that available toxicity data are a good surrogate for FW fish in general.

Comparison of the 10th centile of the distribution of LC_{50}s for arthropods to the measured annual maximum concentrations of chlorpyrifos in the Lake Erie drainage basin suggests that significant effects in some arthropods would be expected to occur. These effects need to be weighed against the transient nature of these responses in the context of the life cycle of these organisms and their resiliency and potential for recovery. In a flowing water system, the distribution of these organisms is not necessarily the same as that of chlorpyrifos, and their generally rapid cycle of reproduction, coupled with repopulation from refugia, makes these groups of organisms less sensitive to higher frequencies of impacts in flowing water systems. This potential for recovery from the impact of chlorpyrifos was demonstrated in the microcosm studies and is supported by similar studies on the impacts of other insecticides in flowing water systems (Kingsbury 1986; Kingsbury and Kreutzweiser 1987; Kreutzweiser and Kingsbury 1987).

E. Population Modeling

To further test the ecological relevance of short-term exposures and the effects of chlorpyrifos on aquatic arthropods, a model of the effects of chlorpyrifos on mortality in populations of *Daphnia magna* was developed using the data from laboratory experiments (Naddy 1996). The model was developed with several objectives: (1) to model responses to exposure scenarios that were not tested in the laboratory, particularly variable concentrations over time; (2) to estimate concentrations at which certain levels of mortality will occur (e.g., LC_{50} and LC_5); and (3) to make long-term predictions, including the time for population recovery and time for a population to reach a certain level.

Model Development. The model structure was based on a population projection matrix approach. Because cladocerans were projected to be among the most sensitive species, even though they generally do not occur in the lesser-order streams where chlorpyrifos concentrations were greatest, *D. magna* was used as a sensitive surrogate species. Furthermore, sufficient information on the effects of chlorpyrifos on age specific survival and fecundity were available (Naddy 1996). In this approach, the population is divided into several cohorts, x, that are projected over time. If $S(x)$ denotes the proportion of the xth cohort ($x = 0,1, \ldots, n$) at the tth point in time ($t = 0,1, \ldots$) that survives to enter the $(x+1)^{st}$ cohort at the $(t+1)^{st}$ time point, the number of individuals which are in the $(x+1)^{st}$ cohort at the tth time point can be estimated (Eq. 10). If the function $f(x)$ denotes the number of offspring born during the interval between the tth and $(t+1)^{st}$ point in time per adult in the xth cohort at the tth time point, the number of individuals in the zeroth cohort can be expressed (equation 11). The population size at time t is the sum of all cohorts at that time (Eq.12).

$$N(x+1), t = S(x)N(x, t-1) \tag{10}$$

$$N(0,t) = \sum_{x=0}^{n} N(x) f(x-1) \tag{11}$$

$$N(t) = \sum_{x=0}^{n} N(x,t) \tag{12}$$

Survival. A model of the time-related mortality response was developed using the principle of reciprocity to account for variable chlorpyrifos exposure concentrations. A graph of the mortality response over time showed a typical sigmoid shape (Naddy 1996) for each exposure concentration. The mortality responses from pulsed exposures, in which mortality lagged behind exposure, was similar to those at constant exposure concentrations. This suggests a threshold exists in exposure beyond which mortality is assured even if exposure ceases. To determine a function to express this process, mortality was plotted against the cumulative exposure concentrations. This relationship also followed a sigmoid response. For each concentration, a logistic function was used to fit the cumulative proportional mortality over the cumulative concentration between cohort birth date bd and an exposure window limit of τ (Eq.13).

$$m(x, C(t)) = \frac{P_1}{1 + e^{(2.2/P_3)(P_2 - \sum_{t=bd}^{\tau} C(t))}} \tag{13}$$

where:
  $m(x, C(t))$ = cumulative proportional mortality of cohort x to τ
  P_1 = maximum cumulative proportional mortality
  P_2 = cumulative concentration yielding peak mortality rate, (d)
  P_3 = cumulative concentration between 0.1 and 0.5 total mortality and between 0.5 and 0.9 total mortality, (d)
  $C(t)$ = chlorpyrifos concentration at time t, (d)

The model was validated by comparing observed and predicted mortality values. The parameters P_1, P_2, and P_3 were similar in magnitude for all exposure scenarios, which confirmed the reciprocity relationships that had been applied to normalize the data to 48-hr LC_{50} values so that they could be compared to 48-hr time-weighted moving averages. The parameters for these functions were estimated using a nonlinear regression procedure (PRISM; GraphPad 1994–1995). The values of P_1 were all set equal to 1.0 and all values of m(t) were scaled between 0 and 1 to form a cumulative probability distribution. To determine the actual mortality, the proportional mortality $m(x,C(t))$ was multiplied by the cohort size $N(x,t)$. To obtain an expression for survival, $S(x)$, the probability of mortality, $m(x,C(t))$, must be subtracted from 1.0.

Reproduction. The birth rate $f(x)$ of a cohort x was estimated using data from experiments in which reproduction in *Daphnia magna* was measured at different chlorpyrifos concentrations (Naddy 1996). No significant difference in young produced per adult was observed among all treatments. The mean and standard deviation of the number of young produced per adult per day, from the age of first reproduction through the end of the 21-d experiments, was 17.6 ± 11.2.

Results of the Simulations. Simulations were run using exposure data from the Lake Erie watershed data set. Results from all simulations predicted exponential growth in the *Daphnia* population (Fig. 29). Although the population was predicted to experience some mortality at the measured exposure concentrations,

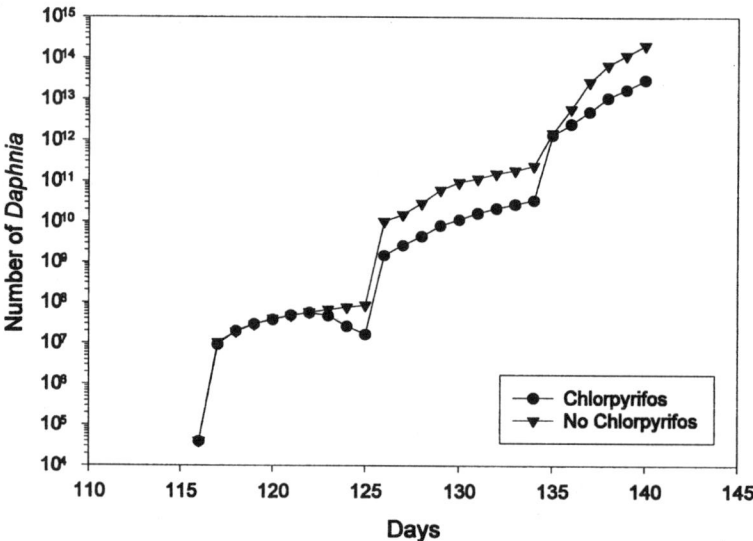

Fig. 29. Population growth of *Daphnia* in the presence and absence of chlorpyrifos exposure.

there were proportions of exposed cohorts that survived to reproduce. Even if a cohort died out, the duration of exposure pulses was short enough and the cumulative concentration sufficiently small to allow other cohorts to survive and continue to reproduce even during the maximum concentration pulse. Estimates of the intrinsic rate of increase for exposed and unexposed populations showed that the growth rate of exposed populations was reduced by only 1.66% at the greatest concentrations of chlorpyrifos observed in streams. Exposure concentrations were multiplied by a constant to predict the concentration at which the population would be reduced by 50%. This would require peak exposures of 450 ng/L and, even under these extreme exposure concentrations, the model predicted population recovery would occur within less than 5 d. This peak concentration is 10 fold greater than the greatest 90th centile chlorpyrifos concentration measured in the Lake Erie drainage basin data set (see Table 11) and would be observed only infrequently. Actual populations would not increase to the levels predicted by the model because of other limiting factors such as predation and density-dependent reductions in productivity. The predicted population levels, however, indicate that *Daphnia* populations should persist and support a normal food web when exposed to chlorpyrifos concentrations as great as those measured in the Lake Erie drainage basin studies.

VII. Uncertainties and Research Needs

The data used in ecological risk assessments are never complete because of the almost infinite variety of circumstances that can exist in natural ecosystems. It can always be argued that additional information would increase the accuracy and reduce the uncertainty of the assessment. Therefore, an assessment of the adequacy of the knowledge base must be made in the context of the assessment endpoints. A great deal is known about the mechanism of toxic action and dose–response relationships for chlorpyrifos. Toxicity information is available for a great number of species from several different classes. The information on toxicity of chlorpyrifos to aquatic organisms was deemed to be adequate to conduct an ecological risk assessment. Many of the studies reviewed were conducted under good laboratory practices (GLP) and a high level of quality assurance and quality control (QA/QC). Information on different life stages of individual species was also available. Furthermore, a great deal of information was available on population- and community-level responses. The only deficiency in the database for aquatic organisms was deemed to be the relatively little information available on the dormant stages of invertebrates or for juvenile and adult amphibians. The few studies available indicated that amphibians are not uniquely sensitive or tolerant in their sensitivity to the effects of chlorpyrifos and that their responses were similar to those of fish. Thus, although this data gap was not judged to be an impediment to conducting the risk assessment, additional information on the exposure and tolerance distributions of amphibians would reduce the uncertainty of the risk assessment.

The quality and quantity of information available on exposure was less well developed than that on toxicity. Sufficient information on the physical, chemical, and environmental properties of chlorpyrifos was available to conduct environmental fate simulations. The greatest limitation was the availability of empirical information on environmental concentrations. Additional information to describe environmental exposures, durations of pulses, and short-term as well as seasonal variation in surface waters would be useful. The data available from the Midwestern corn-growing areas of North America where the most robust data sets were available could be used to focus the study and minimize the analytical needs.

The amount of information on concentrations of chlorpyrifos in the water column of estuaries was limited. While it is believed that dilution results in concentrations of chlorpyrifos in estuaries and coastal marine environments that are unlikely to be sufficiently great to cause adverse effects, because some marine vertebrates were more sensitive than the freshwater vertebrates additional information on these environments would be helpful. The least exposure information was available for bottom sediments in both marine and freshwater environments. The assessment concludes that there is little potential for risk of sediment-dwelling invertebrates in systems where there is little risk to water column organisms. However, because there was little information on concentrations of chlorpyrifos in sediments, this is a very uncertain conclusion. More information on concentrations of chlorpyrifos in sediments may be needed in regions where high water column concentrations have been observed.

An assumption made in the ecological risk assessment presented for chlorpyrifos is that it acts independently of other compounds in the environment. This risk assessment has not specifically considered the potential for chlorpyrifos to act additively or synergistically with other compounds in the aquatic environment. One uncertainty in this assumption is the potential for co-occurrence with other organophosphorus insecticides with the same or similar mode of action. In areas where the risk characterization indicated some potential for adverse effects from exposure of nontarget organisms to chlorpyrifos, it would be useful to refine the risk assessment by considering potential multiple exposures with other organophosphorus insecticides.

Another uncertainty regarding the co-occurrence of toxicants is the potential for synergistic effects. Atrazine has been identified as a possible synergistic substance with chlorpyrifos. Regardless of the applicability of the studies used to demonstrate synergism between atrazine and chlorpyrifos, the assessment error from synergism such as this would only be a factor of approximately twofold. In fact, the potential for synergism would be even less because this conclusion does not consider the probability of co-occurrence of the two chemicals. To reduce this uncertainty would require the investigation of interactions at environmentally relevant concentrations of chlorpyrifos and the potential synergist, atrazine. The appropriate experimental design for these studies would be isobolograms using a range of concentrations of both atrazine and chlorpyrifos in environmentally relevant ratios as well as absolute concentrations. Also, a better description of the probability of co-occurrence would be necessary. The small magnitude of potential supraadditivity does not warrant such studies.

Although this risk assessment was able to identify locations where site-specific risk characterizations should be conducted, it must be recognized that the use of field studies to determine the potential for effects of individual agricultural chemicals on aquatic communities is complicated by the fact that organisms are exposed simultaneously to a number of chemical and nonchemical stressors. As pointed out, it would be difficult, without application of the appropriate experimental design, to separate the effects of one stressor from another. The streams, which were identified as higher risk sites, occur in agricultural regions have been greatly altered from their primordial state. Typically, they are more turbid and warmer than they would have been before alteration of the landscape by agriculture, and the effects of these changes may be more significant that those caused by the presence of pesticides. The use of caged fish and toxicant identification and evaluation (TIE) and toxic units approaches could be applied on a site-specific basis.

Summary

The objective of this risk assessment was to determine the probability and significance of effects of chlorpyrifos [O,O-diethyl O-(3,5,6-trichloro-2-pyridyl) phosphorothioate)]on aquatic ecosystems in North America. The primary focus of the risk assessment was agricultural ecosystems, especially row crops and, in particular, the "corn-belt" agroecosystems. Where data were available, other major crop and noncrop uses of chlorpyrifos were also assessed for their potential contribution to effects. Exposure and effects in freshwater and saltwater environments were considered. Aquatic invertebrates and fish were included in the assessment but amphibians, reptiles, birds, and mammals were not.

The assessment endpoints were selected on the basis of the pathways of chlorpyrifos exposure, patterns of chlorpyrifos toxicity, and judgments about the ecological, economic, and social importance of ecosystem components at risk. The assessment endpoint for fish was population persistence (a function of survival, growth, and recruitment), while for invertebrates it was community productivity.

Measures of effect were survival, growth, or reproduction of individual organisms, as determined in laboratory toxicity tests, and fish survival and invertebrate population abundance in field mesocosm tests. For this risk assessment, acute toxicity values for freshwater and saltwater organisms (fish and invertebrates) constituted the most important set of measurement endpoints. Chronic toxicity, either measured directly or extrapolated from acute toxicity values, was also considered.

The risk assessment used a "lines of evidence" approach and expert judgment to integrate laboratory toxicity measurements and mesocosm results into a reference value for ecological effects of chlorpyrifos. Laboratory toxicity data were used to support a quantitative probabilistic assessment, and the mesocosm studies supported a more qualitative evaluation of ecological significance.

The environmental transformation products of chlorpyrifos are more water soluble than chlorpyrifos, and the environmental fate and significance of the products are dramatically altered because of this difference. Not only are they more easily eliminated by organisms, but they are also less likely to be partitioned into organisms. None of the transformation products of chlorpyrifos is sufficiently toxic or persistent in the environment to be of toxicological concern. Biomagnification of chlorpyrifos or its metabolites in the food chain was judged not to be relevant to the risk assessment. The only molecule judged to be relevant in an ecological risk assessment was the parent molecule, chlorpyrifos.

Estimates of exposure concentrations were developed using two parallel approaches: simulation modeling, and analysis of surface water monitoring data. For the use of chlorpyrifos in corn, spray drift was judged to be a minor route of entry to aquatic systems; the modeling addressed runoff from T-band applications in this use. The modeling effort proceeded through three phases, or tiers. Tier I was a simplistic, generic calculation of concentrations in a pond receiving water running off a generalized agricultural field after a single application of chlorpyrifos, using EPA's GENEEC model. Tier II involved simulation of two representative sites, selected to represent extreme environmental conditions and median values over multiple years, using current renditions of the GLEAMS and EXAMS models. In Tier III, multiple sets of input parameters for GLEAMS and EXAMS were selected from their respective probability distributions, and the model output was, in turn, presented as probability distributions of chlorpyrifos concentrations.

Environmental concentration data for chlorpyrifos in surface waters have been collected from a number of sites in the U.S. These data were the basis for the Tier IV risk assessment, and the sites included Lake Erie, the U.S. Midwest, California, and various agricultural and urban watersheds sampled under the National Water Quality Assessment Program (NAWQA). Sites where monitoring data had been collected coincided with areas identified to be at greater risk for contamination with chlorpyrifos because of magnitude of use and the characteristics of the environment. Thus, the monitoring data were judged to represent areas where greater exposures would likely occur, and risk assessments in these regions would be protective of other areas where use was less prevalent or climatic conditions were such as to minimize runoff or contamination of streams and other bodies of water.

The distribution analysis of 48-hr $LC_{50}/EC_{50}s$ for all freshwater organisms (including rotifers, mollusks, and other insensitive organisms) demonstrated a wide range of sensitivity to chlorpyrifos, spanning five orders of magnitude, and resulted in a 10th centile intercept of 102 ng/L. Freshwater and saltwater vertebrates exhibited a narrower range of sensitivity with 10th centile intercepts of 5358 and 832 ng/L, respectively. Freshwater (FW) and saltwater (SW) arthropods were consistently more susceptible than the vertebrates, and there was a small difference between those species from freshwater and saltwater with 10th centile intercepts of 55 and 15 ng/L, respectively.

Final acute and chronic values (FAVs and FCVs) were calculated using the methods to derive water quality standards prescribed by the Great Lakes Initiative (GLI). Using the four least genus mean acute values (GMAVs), a chlorpyrifos FAV of 148 ng/L was calculated. The mean acute-to-chronic ratio (ACR) for chlorpyrifos in saltwater and freshwater organisms was 8, with an observed range from 1.4 to 181. Application of the average chlorpyrifos ACR of 8 to the FAV of 148 ng/L gave an FCV of 18 ng/L.

Although data on sediment toxicity and concentrations of chlorpyrifos in sediments are sparse, the available data do not suggest that the sediment route of exposure represents a widespread and significant risk. However, where water column concentrations were deemed to be potentially toxic, concentrations in the sediments may also present a risk.

The results of microcosm and mesocosm studies indicate that exposure to chlorpyrifos concentrations <100 ng/L affect few, if any, aquatic taxa. At concentrations of >200 ng/L, effects on invertebrates were more widespread, but nearly all populations recovered within 2–8 wk. At 500 ng/L and greater, survival and growth of some fish species were affected. A concentration of 100 ng/L can be taken as a conservative no-observed-adverse-effect concentration (NOAEC) for overall ecosystem response, and exposures to greater concentrations result in a variety of direct and indirect ecological effects. This NOAEC needs to be judged in the context of the microcosm exposures which were, in many cases, greater than the commonly observed field duration of 48 hr in streams and rivers. Crustaceans (notably cladocerans, copepods, ostracods, and amphipods) and some emergent insects (mayflies) were typically the most sensitive taxa, while other insects, rotifers, and snails were typically more tolerant. Almost invariably, some tolerant taxa increased in abundance when their more sensitive competitors were reduced. Populations of affected species generally recovered rapidly as chlorpyrifos disappeared from the water. The sources for population recovery varied among taxa, but included resistant resting stages, internal refugia, and external ecosystems.

Aquatic ecosystems vary considerably in their sensitivity and response to anthropogenic inputs as the result of differences in use patterns, species composition, flow patterns, geography, climate, and watershed characteristics. It is therefore not the mission of this risk assessment to attempt to force-fit a single acceptable level of effect to all aquatic systems, but rather suggest a maximum allowable toxicant concentration (MATC) for chlorpyrifos to be used as a guide in decision making. This process allows risk assessors to prioritize the potential risks and to eliminate from further consideration species and locations that are unlikely to be at risk. To do this for a particular site, an "exceedence profile" may be constructed by combining the exposure probability distribution with the species impact probability distribution to generate a graphical expression of the percent exceedence of LC_{50}/EC_{50} values against the probability of a species being affected. This joint probability of exposure and response provides an overall estimate of the probability of a particular level of response being achieved at a given location. The probability of exceedence derived from such a plot may then be interpreted in the context of the natural ecology of the potentially affected species, such as habitat requirements, reproduction/development periodicity, and geographic distribution.

The nature of probability distributions is such that the rate of change in exposure concentration to cause a particular response increases asymptotically as the cumulative frequency of including a sensitive species increases. That is, exposure concentrations would have to become infinitely small to include virtually 100% of the possible species. The 10th centile species sensitivity level was chosen, in part, because it lies near the inflection point for many such distributions. In this region, the MATC can be more accurately defined, and there is where the greatest discrimination of concentration and affect can be achieved. The use of the 10th centile of species response in single-species, acute-toxicity tests of lethality was substantiated as the no-observed-effect concentration (NOEC) in field mesocosm studies with chlorpyrifos.

Assessment of the hazards of chlorpyrifos in Tiers I to III predicted unacceptable levels of concern (LOCs) for invertebrates and fish in most of the corn-growing areas. The lack of widespread and repeated reports of fish kills related to use of chlorpyrifos in corn suggested that the modeling approaches and conservative assumptions applied in Tier I to Tier III exposure assessments resulted in overestimates of EECs or that there were other factors mitigating the toxicity of the predicted EECs. Clearly, additional refinement of the exposure assessments was necessary to better understand the potential hazards from chlorpyrifos in surface waters. Using probabilistic approaches, four measures were used to assess the likelihood of exposure exceedences: the freshwater (FW) acute all organism 10th centile (102 ng/L); FW FAV (148 ng/L); the FCV (18 ng/L) derived from FW and saltwater (SW) data; and the FW mesocosm NOAEC (100 ng/L).

The likelihood that either the 10th centile of the acute toxicity distribution or the NOAEC from mesocosm studies would be exceeded in any of the Lake Erie watersheds was small, <10% in all cases. Only in the Huron River was the FCV exceeded for more than 10% of the sampling intervals. This exceedence was determined to be unlikely to be of sufficient duration to cause chronic responses. Exceedences of the annual maximum measured concentrations in the Lake Erie data set were compared to the 10th centile of the toxicity distribution for freshwater fish. In no case was the 10th centile of the fish toxicity distribution exceeded, confirming the judgment that, even over a period of several years, chlorpyrifos concentrations in this drainage basin would not exceed concentrations such that fish might be affected directly. When these annual maxima were compared to the 10th centiles of sensitivity for FW arthropods, they suggested that, in some rivers in this drainage basin, and in some years, significant effects on invertebrates could occur. These infrequent events must be considered in the context of the ability of these groups to recover rapidly from adverse effects.

Assessment of the concentrations of chlorpyrifos in mainstem rivers in California exhibited two trends. There would be only a small probability of exceeding any assessment criteria. However, in some situations, in the smaller tributaries and agricultural drains, there was a greater likelihood that the threshold concentration for protection of invertebrates would be exceeded. The magnitude of these exceedances were less than those at which fish would be at risk from the direct effects of chlorpyrifos. The conclusion that fish would not be adversely affected at most of the likely surface water concentrations of chlorpyrifos is consistent with the fact that few fish kills in the regions studied can be attributed to the use of chlorpyrifos following label instructions. However, fish could be affected indirectly through effects on arthropods that serve as food organisms for fish, particularly young fish. Although such affects would not be expected to occur frequently, field assessments would be necessary to determine the magnitudes of such effects.

Assessment of the NAWQA data set revealed similar trends to the California data with respect to watershed size. The probability of the 10th centile for the acute responses and the mesocosm NOECs being exceeded was <10% in all cases, except for Cherry Creek near Denver, CO. For the rest of the NAWQA data, the months during which greater concentrations occurred in these locations were consistent with use in corn in Nebraska (June) and with use of chlorpyrifos in tree crops and alfalfa in California (March through May). Although there were few data for estuarine systems, probabilities of exceedance were small, and dilution of contaminants is a common feature of these systems. On this basis, the risk to saltwater organisms in these environments was judged to be small. Environmental concentration data specifically associated with termite and turf uses were insufficient to allow a formal assessment of risk; however, the lack of correlation of reports of field incidents and the number of termite applications made on a state basis suggests that this is not a cause of widespread and repeated fish mortality.

Overall, the data on concentrations in freshwater does not suggest ecologically significant risks, except in a few locations. In these cases, site-specific risk assessments should be conducted in the context of the use designations for these systems. The risk assessment also identified several uncertainties that could be reduced by the collection of additional information. In the cases where the Tier IV risk assessment indicates a small but possibly significant overlap of exposure and effect concentrations, it would be prudent to conduct site-specific field studies. This might include more sampling to better describe the frequency distributions of exposure concentrations and durations and site-specific monitoring of benthic invertebrate and fish populations and communities. Also, a combination of caged fish studies and toxicity tests with water or sediments from these locations would decrease the uncertainty in the risk assessment. These types of field or quasi-field studies are complicated by the potential for exposures to multiple toxicants. If effects were observed in field monitoring or semicontrolled or *in situ* bioassays, a toxicant identification and evaluation (TIE) approach would need to be used to assign causality.

Effects of chlorpyrifos on the more sensitive components of biota would not necessarily constitute significant changes in community function in most systems. Assessment criteria that protect all but the most sensitive taxa are also protective of overall ecosystem structure and function. This conclusion is supported by the results from microcosm studies. The risk assessment of chlorpyrifos must be tempered by additional considerations that relate to the spatial and temporal distribution of organisms and the ability of many of the groups of organisms most likely to be affected to recover from quite severe impacts. If site-specific risk assessments and the more intensive, site-specific monitoring of both exposures and population-and community-level responses indicate that adverse effects are being observed under field conditions and these effects can be attributed to exposure to chlorpyrifos, site-specific mitigation measures could be applied to reduce the risk.

Acknowledgments

The authors thank Drs. D.H. Lade, N.N. Poletika, K.B. Woodburn, D.D. Fontaine, M.A. Mayes, and D.A. Laskowski of Dow AgroSciences for providing the resources and technical support that allowed this ecological risk assessment to be conducted.

References

Abernethy SG, Mackay D, McCarty LCS (1988) "Volume fraction" correlation for narcosis in aquatic organisms: the key role of partitioning. Environ Toxicol Chem 7: 469–481.

Anderson PS, Yuhas AL (1996) Improving risk management by characterizing reality: a benefit of probabilistic risk assessment. Hum Ecol Risk Assess 2:55–58.

Ankley GT, Call DJ, Cox JS, Kahl MD, Hoke RA, Kosian PA (1994) Organic carbon partitioning as a basis for predicting the toxicity of chlorpyrifos in sediments. Environ Toxicol Chem 13:621–626.

AQUIRE (1994) Aqua/Info Database 1.4. for PCs. AScI Corp., McLean, VA.

Bailey HC, Miller JL, Miller MJ, Woborg LC, Deanovic L, Shed T (1997) Joint acute toxicity of diazinon and chlorpyrifos to *Ceriodaphnia dubia*. Environ Toxicol Chem 16:2304–2308.

Barron MG, Woodburn KB (1995) Ecotoxicology of chlorpyrifos. Rev Environ Contam Toxicol 144:1–93.

Barron MG, Plakas SM, Wilga PC (1991) Chlorpyrifos pharmacokinetics and metabolism following intravascular and dietary administration in channel catfish. Toxicol Appl Pharmacol 108:474–482.

Barron MG, Plakas SM, Wilga PC, Ball T (1993) Absorption, tissue distribution and metabolism of chlorpyrifos in channel catfish following waterborne exposure. Environ Toxicol Chem 12:1469–1476.

Benke GM, Cheever KL, Mirrer FE, Murphy SD (1974) Comparative toxicity, anticholinesterase action and metabolism of methyl parathion and parathion in sunfish and mice. Toxicol Appl Pharmacol 28:97–109.

Bergamaschi BA, Crepeau KL, Kuivila KM (1997) Pesticides associated with suspended sediments in the San Francisco Bay Estuary, California. Open-File Report 97–24. US Geological Survey, Reston, VA.

Bidlack HD (1976) Degradation of ^{14}C-labeled 3,5,6-trichloro-2-pyridinol in 15 select agricultural soils. Report GH-C 953. DowElanco, Indianapolis, IN.

Biever RC, Giddings JM, Kiamos M, Annunziato MF, Meyerhoff R, Racke K (1994) Effects of chlorpyrifos on aquatic microcosms over a range of off-target spray drift exposure levels. In Proceedings, Brighton Crop Protection Conference on Pests and Diseases, Brighton, UK, November 21–24, 1994, Vol. 3, pp 1367–1372.

Boone JS, Chambers JE (1996) Time course of inhibition of cholinesterase and aliesterase activities, and nonprotein sulfhydryl levels following exposure to organophosphorus insecticides in mosquitofish (*Gambusia affinis*). Fundam Appl Toxicol 29:202–207.

Brazner JC, Kline ER (1990) Effects of chlorpyrifos on the diet and growth of larval fathead minnows, *Pimephales promelas*, in littoral enclosures. Can J Fish Aquat Sci 47:1157–1165.

Brock TCM, Roijackers RMM, Rollon R, Bransen F, Van der Heyden L (1995) Effects of nutrient loading and insecticide application on the ecology of *Elodea*-dominated freshwater microcosms. II. Responses of macrophytes, periphyton and macroinvertebrate grazers. Arch Hydrobiol 134:53–74.

Brock TCM, Crum SJH, van Wijngaarden R, Budde BJ, Tijink J, Zuppelli A, Leeuwangh P (1992) Fate and effects of the insecticide Dursban 4E in indoor *Elodea*-dominated and macrophyte free freshwater model ecosystems: I. Fate and primary effects of the active ingredient chlorpyrifos. Arch Environ Contam Toxicol 23:69–84.

Brown RP, Landre AM, Miller JA, Kirk HD, Hugo JM (1997) Toxicity of sediment-associated chlorpyrifos with the freshwater invertebrates *Hyalella azteca* (amphipod) and *Chironomus tentans* (midge). DECO-ES-3036. Health and Environmental Research Laboratories, Dow Chemical Co., Midland, MI.

Burmaster DE (1996) Benefits and costs of using probabilistic techniques in human health risk assessments—with emphasis on site-specific risk assessments. Hum Ecol Risk Assess 2:35–43.

Burns LA (1990) Exposure Analysis Modeling System: User's Guide for EXAMS II, Version 2.9.4. EPA/600/3-89-084. U.S. Environmental Protection Agency, Washington, DC.

Cairns J Jr (1989) Will the real ecotoxicologist please stand up? Environ Toxicol Chem 8:843–844.

Call DJ, Brooke LT, Knuth ML, Poirier SH, Hoglund MD (1985) Fish subchronic toxicity prediction model for industrial organic chemicals that produce narcosis. Environ Toxicol Chem 4:335–341.

Cardwell RD, Parkhurst BR, Warren-Hicks W, Volosin JS (1993) Aquatic ecological risk. Water Environ Technol 5:47–51.

Carr RL, Chambers JE (1996) Kinetic analysis of the in vitro inhibition, aging, and reactivation of brain acetylcholinesterase from rat and channel catfish by paraoxon and chlorpyrifos-oxon. Toxicol Appl Pharmacol 139:365–373.

Carr RL, Ho LL, Chambers JE (1997) Selective toxicity of chlorpyrifos to several species of fish during an environmental exposure: biochemical mechanisms. Environ Toxicol Chem 16:2369–2374.

Carrington CD (1996) Logical probability and risk assessment. Hum Ecol Risk Assess 2:62–78.

Chakrabarti A, Gennrich SM (1987) Vapor pressure of chlorpyrifos. Report ML-AL-87-40045. DowElanco, Indianapolis, IN.

Chambers JE, Chambers HW (1989) An investigation of acetyl-cholinesterase inhibition and aging and choline acetyltransferase activity following a high level acute exposure to paraoxon. Pestic Biochem Physiol 33:125–131.

Chandler GT, Coull BC, Schizas NV, Donelan T (1997) A culture-based assessment of the effects of chlorpyrifos on multiple meiobenthic copepods using microcosms of intact estuarine sediments. Environ Toxicol Chem 16:2339–2346.

Coppage DL (1972) Organophosphate pesticides: specific level of brain AChE inhibition related to death on sheepshead minnow. Trans Am Fish Soc 101:534–536.

Crum SJH, Brock TCM (1994) Fate of chlorpyrifos in indoor microcosms and outdoor experimental ditches. In: Hill IR, Heimbach F, Leeuwangh P, Matthiessen P (eds) Freshwater Field Tests for Hazard Assessment of Chemicals. Lewis, Boca Raton, FL, pp 315–322.

Cuppen JGM, Glystra R, van Beusekom S, Budde BJ, Brock TCM (1995) Effects of nutrient loading and insecticide application on the ecology of *Elodea*-dominated freshwater microcosms. III. Responses of macroinvertebrate detritivores, breakdown of plant litter, and final conclusions. Arch Hydrobiol 134:157–177.

Daniels RE, Allen JD (1981) Life table evaluation of chronic exposure to a pesticide. Can J Fish Aquat Sci 38: 485–494.

Day K, Kaushik NK (1987) An assessment of the chronic toxicity of the synthetic pyrethroid, fenvalerate, to *Daphnia galeata mendotae*, using life tables. Environ Pollut 44:13–26.

De Bruijn J, Busser F, Seinen W, Hermens J (1989) Determinations of octanol/water partition coefficients for hydrophobic organic chemicals with the "slow-stirring" method. Environ Toxicol Chem 8:499–512.

Deneer JW (1993) Uptake and elimination of chlorpyrifos in the guppy at sublethal and lethal aqueous concentrations. Chemosphere 26:1607–1616.

Deneer JW (1994) Bioconcentration of chlorpyrifos by the three-spined stickleback under laboratory and field conditions. Chemosphere 29:1561–1575.

DiGiorgio C, Bailey HC, Hinton DE (1995) Colorado River Basin toxicity report. Draft final. March 1993–February 1994. Interagency Agreement No. 0–149-250–0. State Water Resources Control Board, Sacramento, CA.

Dilling WL, Lickly LC, Lickly TD, Murphy PG, McKellar RL (1984) Organic photochemistry. 19. Quantum yields for *O,O*-diethyl-*O*-(3,5,6-trichloro-2-pyridyl) phosphorothioate (chlorpyrifos) and 3,5,6-trichloro-2-pyridinol in dilute aqueous solutions and their environmental phototransformation rates. Environ Sci Technol 18:540–543.

DiToro DD, Zarba CS, Hansen DJ, Berry WJ, Swartz RC, Cowan CE, Pavlou SP, Allen HE, Nelson TA, Paquin PR (1991) Technical basis for establishing sediment quality criteria for nonionic organic chemicals using equilibrium partitioning. Environ Toxicol Chem 10:1541–1583.

Domagalski JL, Kuivila KM (1993) Distributions of pesticides and organic contaminants between water and suspended sediment, San Francisco Bay, California. Estuaries 16: 416–426.

Eaton J, Arthur J, Hermanutz R, Kiefer R, Mueller L, Anderson R, Erickson R, Nordling B, Rogers J, Prichard H (1985) Biological effects of continuous and intermittent dosing of outdoor experimental streams with chlorpyrifos. In: Bahner RC, Hansen DJ (eds) Aquatic Toxicology and Hazard Assessment: Eighth Symposium. ASTM STP 891. American Society for Testing and Materials, Philadelphia, PA, pp 85–118.

Environment Canada (1996) Ecological Risk Assessments of Priority Substances under the Canadian Environmental Protection Act: Resource Document Draft 1.0 (March 1996) and Ecological Risk Assessments of Priority Substances under the Canadian Environmental Protection Act: Guidance Manual Draft 2.0 (March 1996). Environment Canada, Ottawa.

ECOFRAM (1997) Joint USEPA, industry and academia workgroup on probabilistic risk assessment for pesticides. January 1997, Washington, D.C. The best reference for ECOFRAM may be the web page www.epa.gov/oppefed1/ecorisk/ as this is the most complete set of information. No hard copies have been published.

Felsot A, Dahm PA (1979) Sorption of organophosphorus and carbamate insecticides by soil. J Agric Food Chem 27:557–563.

Fontaine DD, Wetters JH, Weseloh JW, Stockdale GD, Young JR, Swanson ME (1987) Field dissipation and leaching of chlorpyrifos. Report GH-C 1957. DowElanco, Midland, MI.

Foran JA, Holst LL, Giesy JP (1991). Effects of photoenhanced toxicity of anthracene on ecological and genetic fitness of *Daphnia magna*: a reappraisal. Environ Contam Toxicol 10:425–427.

Friege HL (1986) Monitoring of the River Rhein-experience gathered from accidental events in 1986. In: Organic Micropollution in the Aquatic Environment (CEC 5th European Sympsium, Rome, Italy, October 20–23), pp 132–143.

Giddings JM (1993a) Chlorpyrifos (Lorsban 4E): outdoor aquatic microcosm test for environmental fate and ecological effects. Report 92-6-4288. Springborn Laboratories, Wareham, MA.

Giddings JM (1993b) Chlorpyrifos (Lorsban 4E): outdoor aquatic microcosm test for environmental fate and ecological effects of combinations of spray and slurry treatments. Report 92-11-4486. Springborn Laboratories, Wareham, MA.

Giddings JM, Franco PJ (1985) Calibration of laboratory bioassays with results from microcosms and ponds. In: Boyle TP (ed) Validation and Predictability of Laboratory Methods for Assessing the Fate and Effects of Contaminants in Aquatic Ecosystems. ASTM STP 865. American Society for Testing and Materials, Philadelphia, PA, pp 104–119.

Giddings JM, Biever RC, Racke KD (1997) Fate of chlorpyrifos in outdoor pond microcosms and effects on growth and survival of bluegill sunfish. Environ Toxicol Chem 16:2353–2362.

Giesy JP, Graney RL (1989) Recent developments in and intercomparisons of acute and chronic bioassays. Hydrobiologia 188/189:21–60.

Gorzinski SJ, Mayes MA, Ormand JR, Weinberg JT, Richardson CH (1991a) 3,5,6-Trichloro-2-pyridinol: acute 96-hr toxicity to the bluegill, *Lepomis macrochirus* Refinesque. Report ES-DR-0037–0423-7. DowElanco, Midland, MI.

Gorzinski SJ, Mayes MA, Ormand JR, Weinberg JT, Richardson CH (1991b) 3,5,6-trichloro-2-pyridinol: Acute 96-hr toxicity to the rainbow trout (*Oncorhynchus mykiss* Walbaum). Report ES-DR-0037–0423-7. DownElanco, Midland, MI.

Gorzinski SJ, Mayes MA, Ormand JR, Weinberg JT, Richardson CH (1991c) 3,5,6-Trichloro-2-pyridinol: acute 96-hr toxicity to the water flea (*Daphnia magna* Straus). Report ES-DR-0037–0423-7. DowElanco, Midland, MI.

Graney RL, Kennedy JH, Rodgers JH (eds.) (1994) Aquatic Mesocosm Studies in Ecological Risk Assessment. CRC Press, Boca Raton, FL.

GraphPad (1994–1995) GraphPad PRISM Version 2.0. GraphPad Software, San Diego, CA.

Graves WC, Smith GJ (1991a) 3,5,6-Trichloro-2-pyridinol: a 96-hour flow-through acute toxicity with the grass shrimp (*Palamonetes pugio*). Report DECO-ES-2457. DowElanco, Midland, MI.

Graves WC, Smith GJ (1991b) 3,5,6-Trichloro-2-pyridinol: a 96-hour flow-through acute toxicity test with the Eastern oyster (*Crassostrea virginica*). Report DECO-ES-2548. DowElanco, Midland, MI.

Green AS, Chandler GT (1996) Life-table evaluation of sediment-associated chloropyrifos chronic toxicity to the benthic copepod, *Amphiascus tenuiremis*. Arch Environ Contam Toxicol 31:77–83.

Hall LW, Anderson RD (1993) The influence of salinity on the toxicity of various classes of chemicals to aquatic biota. Crit Rev Toxicol 25:281–346.

Hansen SC, Woodburn KB, Wilga PC, Ball T (1992) Chlorpyrifos: distribution and metabolism in the eastern oyster, *Crassostrea virginica*. Report DECO-ES-2377. Dow Chemical Company, Midland, MI.

Hasspieler BM, Behar JV, DiGiulio RT (1994) Glutathione-dependent defense in channel catfish (*Ictalurus punctatus*) and brown bullhead (*Ameriurus nebulosus*). Ecotoxicol Environ Saf 28:82–90.

Havens PL, Helly JJ, Mangels G, Parker RD (1995) Establishing scientific criteria to determine the probability of pesticide runoff. Gather/Scatter (October-December 1995), pp 8–9.

Havens PL, Cryer SA, Rolston LJ (1998) Tiered aquatic risk refinement: case study—at-plant applications of granular chlorpyrifos to corn. Environ Toxicol Chem 17:1313–1322.

Health Council of the Netherlands (1993) Ecotoxicological risk assessment and policy-making in the Netherlands—dealing with uncertainties. Network 6(3)/7(1):8–11.

Hedlund RT (1973) Bioconcentration of chlorpyrifos by mosquito fish in a flowing system. Report GS-1318. Dow Chemical Compay, Midland, MI.

Hoke RA, Ankley GT, Cotter AM, Goldstein PA, Kosian PA, Phipps Gl, Vandermeiden FM (1994) Evaluation of equilibrium partitioning theory for predicting acute toxicity of field-collected sediments contaminated with DDT, DDE and DDD to the amphipod *Hyalella azteca*. Environ Toxicol Chem 13:157–166.

Holcombe GW, Phipps GL, Tanner DK (1982) The acute toxicity of kelthane, dursban, disulfoton, pydrin, and permethrin to fathead minnows *Pimephales promelas* and rainbow trout *Salmo gairdneri*. Environ Pollut 29:167–178.

Hurlbert SH, Mulla MS, Willson HR (1972) Effects of an organophosphorus insecticide on the phytoplankton, zooplankton, and insect populations of fresh-water ponds. Ecol Monogr 42:269–299.

Hurlbert SH, Mulla MS, Keith JO, Westlake WE, Dhsch ME (1970) Biological effects and persistence of Dursban in freshwater ponds. J Econ Entomol 63:43–52.

Jarvinen AW, Tanner DK (1982) Toxicity of selected controlled release and corresponding unformulated technical grade pesticides to the fathead minnow *Pimephales promelas*. Environ Pollut Series A 27:179–195.

Jarvinen AW, Tanner DK, Kline ER (1988) Toxicity of chlorpyrifos, endrin, or fenvalerate to fathead minnows following episodic or continuous exposure. Ecotoxicol Environ Saf 15:78–95.

Johnson JA, Wallace KB (1987) Species-related differences in the inhibition of brain acetylcholinesterase by paraoxon and malaoxon. Toxicol Appl Pharmacol 88:234–241.

Kaushik, NK, Solomon, KR, Stephenson, GL and Day KE. 1986. Use of limnocorrals in evaluating effects of pesticides on zooplankton communities pp. 269–290. In community toxicity testing. John Cairns (Ed.) ASTM STP 920. American Society of Testing and Materials. Philadelphia, PA.

Kenaga EE (1982) Predictability of chronic toxicity from acute toxicity of chemicals in fish and aquatic invertebrates. Environ Toxicol Chem 1:347–358.

Kingsbury PD (1986) Effects of an aerial application of the pyrethroid permethrin on a forest stream. Manit Entomol 10:9–17.

Kingsbury PD, Kreutzweiser DP (1987) Permethrin treatments in Canadian forests. Part I: Impact on stream fish. Pestic Sci 19:35–48.

Klaine SJ, Cobb GP, Dickerson RL, Dixon KR, Kendall RJ, Smith EE, Solomon KR (1996) An ecological risk assessment for the use of the biocide, dibromonitrilopropionamide (DBNPA) in industrial cooling systems. Environ Toxicol Chem 15:21–30.

Knuth ML, Heinis LJ (1992) Dissipation and persistence of chlorpyrifos within littoral enclosures. J Agric Food Chem 40:1257–1263.

Kreutzweiser DP, Kingsbury PD (1987) Permethrin treatments in Canadian forests. Part II: Impact on stream invertebrates. Pestic Sci 19:49–60.

Leahy PP, Rosenshein JS, Knopman DS (1990) Implementation plan for the National Water-Quality Assessment Program. Open-File Report 90–174. U.S. Geological Survey, Reston, VA.

Leeuwangh P (1994) Comparison of chlorpyrifos fate and effects in outdoor aquatic micro-and mesocosms of various scale and construction. In: Hill IR, Heimbach F, Leeuwangh P, Matthiessen P (eds) Freshwater Field Tests for Hazard Assessment of Chemicals. Lewis, Boca Raton, FL, pp 217–248.

Leeuwangh P, Brock TCM, Kersting K (1994) An evaluation of four types of freshwater model ecosystem for assessing the hazard of pesticides. Hum Exp Toxicol 13:888–899.

Lucassen WGH, Leeuwangh P (1994) Response of zooplankton to Dursban 4E insecticide in a pond experiment. In: Graney RL, Kennedy JH, Rodgers JH (eds) Aquatic Mesocosm Studies in Ecological Risk Assessment. Lewis, Boca Raton, FL, pp 517–533.

MacCoy D, Crepeau KL, Kuivala KM (1995) Dissolved pesticide data for the San Joaquin River at Vernalis and the Sacramento River at Sacramento, California, 1991–1994. Open-File Report 95-110. U.S. Geological Survey, Sacramento, CA.

Macek KJ, Walsh DF, Hogan JW, Holz DD (1972) Toxicity of the insecticide Dursban to fish and aquatic invertebrates in ponds. Trans Am Fish Soc 101:420–427.

Mansingh A, Robinson DE, Dalip KM (1997) Insecticide contamination of the Jamaican environment. Trends Anal Chem 16:115–123.

Marshall WK, Roberts JR (1978) Ecotoxicology of chlorpyrifos. NRCC 16079. National Research Council of Canada, Ottawa, Ontario.

McCall PJ (1987) Soil adsorption properties of ^{14}C-chlorpyrifos. Report GH-C 1971. DowElanco, Midland, MI.

McCarty LS (1986) The relationship between aquatic toxicity QSARs and bioconcentration for some organic chemicals. Environ Toxicol Chem 5:1071–1080.

McConnell LL, Nelson E, Rice CP, Baker JE, Johnson WE, Harman JA, Bialek K (1997) Chlorpyrifos in the air and surface water of Chesapeake Bay: predictions of atmospheric deposition fluxes. Environ Sci Technol 31:1390–1398.

McDonald IA, Howes DA, Gillis NA (1985) The determination of the physico-chemical parameters of chlorpyrifos. Report GH-C 1393. DowElanco, Midland, MI.

Metcalf RL (1974) A laboratory model ecosystem to evaluate compounds producing biological magnification. Essays Toxicol 5:17–28.

Meyer JS, Ingersoll CG, McDonald LL (1987) Sensitivity analysis of population growth rates estimated from cladoceran toxicity tests. Arch Environ Contam Toxicol 6:115–126.

Montague B (1996) Pesticide Toxicity Database. USEPA database of pesticide toxicity information. Office of Pesticide Programs, U.S. Environmental Protection Agency, Washington, DC.

Montañés JF, Van Hattum B, Deneer J (1995) Bioconcentration of chlorpyrifos by the freshwater isopod *Asellus aquaticus* (L.) in outdoor experimental ditches. Environ Pollut 88:137–146.

Motoyama N, Dauterman WC (1980) Glutathione S-transferases: their role in the metabolism of organophosphorus insecticides. In: Hodgson E, Bend JR, Philpot RM (eds) Rev Biochem Toxicol 2:49–69.

Mullins JA, Carsel RF, Scarbrough JE, Ivery AM (1993) PRZM-2: a model for predicting pesticide fate in the crop root zone and unsaturated soil zones: program and user's manual for release 2.0. EPA/600/R-93/046. U.S. Environmental Protection Agency, Athens, GA.

Murphy PG, Lutenski NE (1986) Bioconcentration of chlorpyrifos in rainbow trout *(Salmo gairdneri* Richardson). DowElanco, Indianapolis, IN.

Naddy RB (1996) Assessing the toxicity of the organophosphorus insecticide chlorpyrifos to a freshwater invertebrate: *Daphnia magna* (Crustacea: Cladocera). Ph.D. dissertation, Clemson University, Clemson, SC.

Neely WB, Branson DR, Blau GE (1974) Partition coefficient to measure bioconcentration potential of organic chemicals in fish. Environ Sci Technol 8:1113–1115.

Nicks A (1989) CLIGEN weather generator program (Version 1.0). USDA-ARS WEPP Water Erosion Project, Durant, OK.

NRC (1993) Issues in Risk Assessment. National Research Council, National Academy Press, Washington, DC.

Packard SR (1987) Determination of the water solubility of chlorpyrifos. Report ML-AL 87-71102. DowElanco, Midland, MI.

Pait AS, DeSouza AE, Farrow RH (1992) Agricultural pesticide use in coastal areas: a national survey. Final report, National Oceanic and Atmospheric Administration, Rockville, MD.

Pape-Lindstrom PA, Lydy MJ (1997) Synergistic toxicity of atrazine and organophosphate insecticides contravenes the response addition mixture model. Environ Toxicol Chem 16:2415–2420.

Parker RD, Rieder DD (1995) The generic expected environmental concentration program, GENEEC. Part B, users manual, tier one screening model for aquatic pesticide exposure. Environmental Fate and Effects Division, Office of Pesticide Programs, USEPA,Washington, DC.

Parkhurst BR, Warren-Hicks W, Etchison T, Butcher JB, Cardwell RD, Volison J (1995) Methodology for aquatic ecological risk assessment. RP91-AER. Final report prepared for the Water Environment Research Foundation, Alexandria, VA.

Pereira WE, Domagalski JL, Hostettler FD, Brown LR, Rapp JB (1996) Occurrence and accumulation of pesticides and organic contaminants in river sediment, water and clam tissues from the San Joaquin River and tributaries, California. Environ Toxicol Chem 15:172–180.

Phipps GL, Holcombe GW (1985) A method for aquatic multiple species toxicant test-

ing: acute toxicity of 10 chemicals to 5 vertebrates and 2 invertebrates. Environ Pollut Series A 38:141–157.
Power M, McCarty LS (1996) Probabilistic risk assessment: betting on its future. Hum Ecol Risk Assess 2:30–34.
Racke KD (1993) Environmental fate of chlorpyrifos. Rev Environ Contam Toxicol 131: 1–154.
Racke KD, Robb CK (1993) Dissipation of chlorpyrifos in warm-season turfgrass and fallow soil in Florida. Report GH-C 3077. DowElanco, Midland, MI.
Richards RP, Baker DB (1993) Pesticide concentration patterns in agricultural drainage networks in the Lake Erie Basin. Environ Toxicol Chem 12:13–26.
Richardson GM (1996) Deterministic versus probabilistic risk assessment: strengths and weaknesses in a regulatory context. Hum Ecol Risk Assess 2:44–54.
Rice PJ (1992) Acute toxicity and behavioral effects of chlorpyrifos, peumethrin, phend, strychnine and 2,4-dini Trophenol to 30-day old Japanese medaka (*ORYZIAS LATIPES*) Environ Toxicol Chem 16:696–704.
Ross L (1992a) Preliminary results of the San Joaquin River study; summer 1991. Staff memorandum, May 21, 1992. California Department of Pesticide Regulation, Sacramento, CA.
Ross L (1992b) Preliminary results of the San Joaquin River study: winter 1991–1992. Staff memorandum, May 22, 1992. California Department of Pesticide Regulation, Sacramento, CA.
Ross L (1993a) Preliminary results of the San Joaquin River study: spring 1992. Staff memorandum, January 29, 1993. California Department of Pesticide Regulation, Sacramento, CA.
Ross L (1993b) Preliminary results of the San Joaquin River study; summer 1992. Staff memorandum, September 22, 1993. California Department of Pesticide Regulation, Sacramento, CA.
Ross L (1993c) Preliminary results of the San Joaquin River study; winter 1992–1993. Staff memorandum, September 23, 1993. California Department of Pesticide Regulation, Sacramento, CA.
Schimmel SC, Garnas RL, Patrick JM, Moore JC (1983). Acute toxicity, bioconcentration and persistence of AC 222,705, benthiocarb, chlorpyrifos, fenvalerate, methyl parathion, and permethrin in the estuarine environment. J Agric Food Chem 31:104–113.
SETAC (Society of Environmental Toxicology and Chemistry)(1994) Pesticide risk and mitigation. Final Report of the Aquatic Risk Assessment and Mitigation Dialog Group. SETAC Foundation for Environmental Education, Pensacola, FL.
Siefert RE (1984) Effects of Dursban (chlorpyrifos) on non-target aquatic organisms in a natural pond undergoing mosquito control treatment. Progress Report. USEPA, Duluth, MN.
Siefert RE, Brazner JC, Knuth ML, Heinis LJ, Jensen DA, Larson N (1987) Effects of Dursban (chlorpyrifos) on aquatic organisms in enclosures in a natural pond. Final Report. USEPA, Environmental Research Laboratory, Duluth, MN.
Siefert RE, Lozano SJ, Brazner JC, Knuth ML (1989) Littoral enclosures for aquatic field testing of pesticides: effects of chlorpyrifos on a natural system. In Voshell JR Jr (ed) Using Mesocosms to Assess the Aquatic Ecological Risk of Pesticides: Theory and Practice. Miscellaneous Publication 75. Entomological Society of America, Lanham, MD, pp 57–73.
SigmaPlot (1997) SPSS Inc. Version 4 for Windows 95, SPSS Inc., 444 N. Michigan Ave., Chicago, IL.

Smith GN, Watson BS, Fischer FS (1966) The metabolism of [^{14}C]O,O-diethyl-O-(3,5,6-trichloro-2-pyridyl) phosphorothioate (Dursban) in fish. J Econ Entomol 59:1464–1475.

Solomon KR (1996) Overview of recent developments in ecotoxicological risk assessment. Risk Anal 16:627–633.

Solomon KR, Chappel MJ (1998) Triazine herbicides: ecological risk assessment in surface waters. In: Ballantine L, McFarland J, Hackett D (eds) Triazine Risk Assessment. A.C.S. Symposium Series Vol. 683. American Chemical Society, Washington DC, pp 357–368.

Solomon KR, Baker DB, Richards P, Dixon KR, Klaine SJ, La Point TW, Kendall RJ, Giddings JM, Giesy JP, Hall LW Jr, Weisskopf CP, Williams M (1996) Ecological risk assessment of atrazine in North American surface waters. Environ Toxicol Chem 15:31–76.

Stephenson GL, Kaushik NK, Solomon KR, Day KE, Hamilton P (1986) Impact of methoxychlor on freshwater plankton communities in limnocorals. Environ Toxicol Chem 5:587–603.

Somasundaram L, Coats JR, Shanbhag VM, Stahr HM (1991) Mobility of pesticides and their hydrolysis metabolites in soil. Environ Toxicol Chem 10:185–194.

Suter G II, Barnthouse LW, Bartell SM, Mill T, Mackay D, Patterson S (1993) Ecological Risk Assessment. Lewis, Boca Raton, FL.

Thomas JD, Chambers JE (1996) A retrospective analysis of surface water contamination by chlorpyrifos-based termiteicide emulsions (Dursban* TC, Equity Termiticide) based on water incident survey and analytical data. DowElanco, Indianapolis, IN.

Urban DJ, Cook NJ (1986) Hazard Evaluation Division standard evaluation procedure ecological risk assessment. EPA-540/9-85-001. USEPA, Washington, DC.

USEPA (U.S. Environmental Protection Agency) (1986) Ambient water quality criteria for chlorpyrifos—1986. EPA 440/5-86-005. Office of Water, Washington, DC.

USEPA (U.S. Environmental Protection Agency) (1992a) Framework for ecological risk assessment. EPA/630/R-92/001. USEPA, Washington, DC.

USEPA (U.S. Environmental Protection Agency) (1992b) National study of chemical residues in fish. EPA/823/R/92/008a. Office of Science and Technology (WH-551), Washington, DC.

USEPA (U.S. Environmental Protection Agency) (1995) The use of the benchmark dose approach in health risk assessment. Risk assessment forum. EPA/630/R-94/007. USEPA, Washington, DC.

USEPA (U.S. Environmental Protection Agency) (1996a) Proposed guidelines for ecological risk assessment; notice. Fed Reg 61:47552–47631.

USEPA (U.S. Environmental Protection Agency) (1996b) Draft corn insecticide cluster analysis. Environmental Fate and Effects Division, Office of Pesticide Programs, Washington, DC.

U.S. Department of Agriculture (USDA) (1992) Groundwater loading effects of agriculture management systems (GLEAMS), model Version 2.10. Soil Conservation Section, USDA, Washington, DC.

van den Brink PJ, van Donk E, Gylstra R, Crum SJH, Brock TCM (1995) Effects of chronic low concentrations of the pesticides chlorpyrifos and atrazine in indoor freshwater microcosms. Chemosphere 31:3181–3200.

van den Brink PJ, van Wijngaarden RPA, Lucassen WGH, Brock TCM, Leeuwangh P (1996) Effects of the insecticide Dursban 4E (active ingredient chlorpyrifos) in outdoor experimental ditches: II. Invertebrate community responses and recovery. Environ Toxicol Chem 15:1143–1153.

van Donk E, Prins H, Voogd HM, Crum SJH, Brock TCM (1995) Effects of nutrient loading and insecticide application on the ecology of *Elodea*-dominated freshwater microcosms. I. Responses of plankton and zooplanktivorous insects. Arch Hydrobiol 133:417–439.

van Wijngaarden RPA, Leeuwangh P (1993) Relationship between toxicity in laboratory and pond: an ecotoxicological study with chlorpyrifos. Meded Fac Landbouwwet Rijksun Gent 54:1061–1069.

van Wijngaarden RPA, Leeuwangh P, Lucassen WGH, Romijn K, Ronday R, van der Velde R, Willigenburg W (1993) Acute toxicity of chlorpyrifos to fish, a newt, and aquatic invertebrates. Bull Environ Contam Toxicol 51:716–723.

van Wijngaarden RPA, van den Brink PJ, Crum SJH, Voshaar JHO, Brock TCM, Leeuwangh P (1996) Effects of the insecticide Dursban 4E (active ingredient chlorpyrifos) in outdoor experimental ditches: I. Comparison of short-term toxicity between the laboratory and the field. Environ Toxicol Chem 15:1133–1142.

Wallace KB, Herzberg U (1988) Reactivation and aging of phosphorylated brain acetylcholinesterase from fish and rodents. Toxicol Appl Pharmacol 92:307–314.

Walthall WK, Stark JD (1997) A comparison of acute mortality and population growth rate as endpoints of toxicological effect. Ecotoxicol Environ Saf 37:45–57.

Wan MT, Moul DJ, Watts RG (1987) Acute toxicity to juvenile Pacific salmonids of Garlon 3A, Garlon 4, triclopyr, triclopyr ester, and their transformation products: 3,5,6-trichloro-2-pyridinol and 2-methoxy-3,5,6-trichloropyridine. Bull Environ Contam Toxicol 39:721–728.

Weiss CM (1961) Physiological effect of organic phosphorus insecticides on several species of fish. Trans Am Fish Soc 90:143–152.

Welling W, de Vries JW (1992) Bioconcentration kinetics of the organophosphorous insecticide chlorpyrifos in guppies (*Poecilia reticulata*). Ecotoxicol Environ Saf 23:64–75.

Zaugg SD, Sandstrom MW, Smith SG, Fehlberg KM (1995) Methods of analysis by the U.S. Geological Survey National Water Quality Laboratory: determination of pesticides in water by C-18 solid-phase extraction and capillary-column gas chromatography/mass spectrometry with selected-ion monitoring. Open-File Report 95–181. U.S. Geological Survey, Denver, CO.

Manuscript received May 27, 1998; accepted June 8, 1998.

Cumulative and Comprehensive Subject Matter Index Volumes 151–160

α-HCH, residues farm products India, **152**:14
α$_w$, absorption efficiency metals marine mussels, **151**:48
Abbreviations, chemical warfare agents, **156**:152
Acetylcholinesterase (AChE) inhibition, nerve agents, **156**:18
Acetylthiocholine iodide, RBC-ChE activity, **156**:22
AChE, chlorpyrifos phosphorylated (diag), **160**:13
AChE (acetylcholinesterase) inhibition, nerve agents, **156**:18
Acidovorax facilis, PHA biodegradation, **159**:17
Acinetobacter, as opportunistic pathogen, **152**:61
Acinetobacter, meningitis causal pathogen, **152**:71
Acinetobacter, occurrence mineral water sources, **152**:61
Acinetobacter, opportunistic pathogens drinking water, **152**:58
Acinetobacter, pneumonia causal pathogen, **152**:71
Acinetobacter, septicemia causal pathogen, **152**:72
Acipenser transmontanus, white sturgeon, **159**:98
Acquired immunodeficiency syndrome (AIDS), opportunistic bacteria water, **152**:58
Acronyms, chemical warfare agents, **156**:152
Activated sewage sludge, alkyl halide probe responses, **155**:59
Acute inhalation toxicity, CK, **156**:114
Acute toxicity, CK (cyanogen chloride), **156**:113
Acute toxicity, GA, **156**:68

Acute toxicity, GB, **156**:78
Acute toxicity, GD, **156**:94
Acute toxicity, HD, **156**:30
Acute toxicity, HN2, **156**:51
Acute toxicity, HT, **156**:49
Acute toxicity, lewisite, **156**:103
Acute toxicity, PCDEs trout, **157**:139
Acute toxicity, T sulfur mustard, **156**:50
Acute toxicity, VX, **156**:54
ADI (average daily intake), organochlorine pesticides various countries, **152**:36
Aeromonas, as diarrheal agent, **152**:63
Aeromonas caviae, as opportunistic pathogen, **152**:63
Aeromonas hydrophila, as opportunistic pathogen, **152**:63
Aeromonas hydrophila, infection risk drinking water, **152**:74
Aeromonas hydrophila, meat irradiation sensitivity, **154**:24, 38
Aeromonas, occurrence in drinking water, **152**:62, 65
Aeromonas, opportunistic pathogens drinking water, **152**:58
Aeromonas, oral infective dose humans, **152**:64
Aeromonas, pneumonia causal pathogen, **152**:71
Aeromonas sobria, as opportunistic pathogen, **152**:63
Africa, organochlorine insecticide use, **151**:2
Africa, organochlorine residues fauna, **151**:1 ff.
Africa, pesticide import values, **151**:4
African fauna, organochlorine insecticide residues, **151**:1 ff.
African restrictions, organochlorine insecticide use, **151**:3

Agent CK, see CK or cyanogen chloride, **156**:9
Agent GA, see GA or Tabun, **156**:9
Agent GB, see GB or Sarin, **156**:9
Agent GD, see GD or Soman, **156**:9
Agent HD, see HD or Sulfur mustard HD, **156**:9
Agent HN2, see HN2 or Nitrogen mustard, **156**:9
Agent HT, see HT or Sulfur mustard HT, **156**:9
Agent L, see L or Lewsite, **156**:9
Agent T, see T Sulfur mustard, **156**:9
Agent VX, see VX, **156**:9
Agrobacterium tumefaciens, "Trojan horse" bacterium, **159**:14
AIDS (acquired immunodeficiency syndrome), opportunistic bacteria water, **152**:58
Air pollutants, plant biomonitors, **157**:1 ff.
Air pollution biomonitors, plant taxa choice, **157**:13
Air pollution biomonitors, plants *vs* animals, **157**:16
Air pollution, plants as biomonitors, **157**:1 ff.
Air-water interface, natural organic matter effects, **155**:122
Airborne herbicide residues, ethalfluralin, **153**:73
Airborne herbicide residues, trifluralin, **153**:40
Airborne tire fragments, health effects, **151**:81
Alcaligenes eutrophus, PHB degradation pathway, **159**:18
Alcaligenes faecalis, PHA biodegradation, **159**:17
Alcaligenes spp, PHA accumulation, **159**:7
Alcaligines, radiation/temperature sensitivity, **154**:10
Aldrin/dieldrin, ADIs various countries, **152**:36
Aldrin, residues food India, **152**:15
Aldrin, residues food Thailand, **152**:22
Aliesterases (carboxylesterases), nerve agent effects, **156**:21

Alkanes, polychlorinated, environmental chemistry/toxicology, **158**:53 ff.
Alkyl halide probe responses, activated sewage sludge, **155**:59
Alkyl halide probe responses, dumpsite soil, **155**:56
Alkyl halide probe responses, lake sediments, **155**:57
Alkyl halide probe responses, marine coastal sediments, **155**:58
Alkyl halide reduction, chromous sulfate, **155**:21
Alkyl halide reduction, iron(II) deuteroporphyrin IX, **155**:26
Alkyl halides, biochemical hydrolysis, **155**:45
Alkyl halides, dehalogenation site reactivity probes, **155**:49
Alkyl halides, fell protein oxidation rates, **155**:29
Alkyl halides, microbial hydrogenolysis, **155**:50
Alkyl halides, microbial hydrolysis, **155**:45
Alkyl halides, microbial oxygen insertion, **155**:49
Alkyl halides, microbial reductive elimination, **155**:50
Alkyl halides, microbial substitution, **155**:48
Alkyl halides, natural synthesis, **155**:3
Alkyl halides, soil bacteria transformations, **155**:47
Allolobophora chlorotica, dimethoate effects, **154**:95
Aluminum, bioaccumulation freshwater insect larvae, **158**:132
Aluminum, contamination Rio Grande Basin biota, **158**:10
Aluminum, trace contamination estuaries, **155**:73 ff.
Amazon, mercury contamination, **157**:25 ff.
American mink (*Mustela vison*), PCB effects, **157**:107
American River otter (*Lutra canadensis*), PCB levels, **157**:105
American River otter vs Eurasian otter, PCB levels, **157**:105

Americium, assimilation efficiency marine mussels, **151**:44
Ames test for mutagenicity/carcinogenicity, **158**:114
Ames test, mutagenic effects rubber accelerators, **151**:82
Ammeline, atrazine metabolite, **151**:120
Amphibians, DDE residues Rio Grande Basin, **158**:20
Amphibians, organochlorine residues Africa, **151**:8, 10
Amphibians, Rio Grande Basin, **158**:5
Amphibians, scientific names (list), **158**:5
Ampicillin, related to drinking water pathogen infection, **152**:74
Analytical methods, blood lead, **159**:30
Analytical methods, cadmium, **154**:70, 72
Analytical methods, EDTA/DTPA, **152**:86
Analytical methods, PCBs in otters, **157**:104
Analytical methods, PCDEs marine samples, **157**:134
Analytical methods, plant atmospheric biomonitors, **157**:8
Analytical methods, polychlorinated alkanes, **158**:71
Analytical methods, sediment-trace metal sample analysis, **155**:88, 91
Analytical methods, trifluralin environmental samples, **153**:62
Analytical procedure, PCDEs (illus.), **157**:135
Animals, Rio Grande Basin, **158**:5
Animals, scientific names (list), **158**:5
Anisakis spp., irradiation control seafood, **154**:40
ANSWERS, pesticide runoff simulation model, **153**:23
Antibiotics, related to drinking water pathogen infection, **152**:74
Anticholinesterase agents, **156**:18
Antimony, bioaccumulation freshwater insect larvae, **158**:133
Antioxidants, rubber tire manufacture, **151**:70, **151**:74
Appliance use, child brain tumors, **159**:124, 126

Appliance use, prenatal exposure child brain tumors, **159**:125
Aquatic animals, mercury biomonitors, **157**:35
Aquatic ecosystems, pesticide effects, **159**:95 ff.
Aquatic environment, heavy metals effects (illus.), **157**:57
Aquatic environment, mercury contamination Amazon, **157**:34
Aquatic environments, chlorpyrifos risk assessment, **160**:1 ff.
Aquatic food chain, DDT Africa chart, **151**:29
Aquatic organisms, EDTA/DTPA effects, **152**:96
Aquatic organisms, polychlorinated alkanes contamination, **158**:77, 83
Aquatic plants, mercury biomonitors, **157**:34
Aquatic systems, sediment pollution assessment, **159**:44
Archebacteria, F-430-containing, **155**:43
Army chemical destruction, **156**:3 ff.
Army chemical non-stockpile sites U.S., **156**:4
Arsenic, bioaccumulation freshwater insect larvae, **158**:132
Arsenic, contamination Rio Grande Basin biota, **158**:10
Arsenic, shark levels Gulf of Mexico, **157**:78
Arsenic, trace contamination estuaries, **155**:73 ff.
Artificial electron acceptors, dehydrogenase activity assays, **159**:60
Asparagopsis taxiformis, haloorganics contained, **155**:3
Aspergillus fumigatus, PHA biodegradation, **159**:17
Aspergillus spp., spices irradiation control, **154**:19
Aspergillus terreus, radiation sensitivity food, **154**:8
Assimilation efficiency, metal bioavailability aquatic organisms, **151**:40
Astrocytoma, child brain tumors, **159**:112
Atmospheric biomonitors, indicator gardens, **157**:6

Atmospheric emission monitoring, plant biomonitors, **157**:10
Atmospheric lead, from lead oxide/alkyl lead, **159**:26
Atmospheric oxidation, halocarbons, **155**:9
Atmospheric pollutants, biomonitors plants vs animals, **157**:16
Atmospheric pollutants, effects plant growth, **157**:11
Atmospheric pollutants, interaction plant biomonitors, **157**:12
Atmospheric pollution bioindicator, defined, **157**:3
Atmospheric pollution, plant biomonitors, **157**:1 ff.
Atmospheric pollution regulation, plant biomonitors, **157**:1 ff.
Atmospheric pollution, selection plant biomonitors, **157**:7
Atomic absorption spectrophotometry, plant atmospheric biomonitors, **157**:8
Atomic fluorescence spectrometry, rubber leachates, **151**:76
ATP assays, microbial sediment toxicity testing, **159**:57
Atrazine, annual use in U.S., **151**:118
Atrazine, annual use on corn, **151**:118
Atrazine, appearance rainwater, **151**:118
Atrazine, application rates, **151**:118
Atrazine, bound residues soil, **151**:130
Atrazine, buffer zones runoff effects, **151**:152
Atrazine, clay sorption, **151**:127
Atrazine, complete degradation to CO_2, **151**:122
Atrazine, controlled-release formulation, **151**:151
Atrazine, cultural management practices, **151**:149
Atrazine decomposition, abiotic, **151**:121
Atrazine decomposition, biotic, **151**:122
Atrazine, degradation in aquifers, **151**:124
Atrazine, degradation pathway, **151**:121
Atrazine, dissolved organic carbon vs soil binding, **151**:131
Atrazine, fate North Central U.S. soils, **151**:117 ff.
Atrazine, field persistence, **151**:145

Atrazine, Freundlich isotherms clay adsorption, **151**:127
Atrazine, groundwater contamination, **151**:118
Atrazine, groundwater correlation with NO_3^-, **151**:136
Atrazine, Henry's Law Constant, **151**:133
Atrazine herbicide, description, **151**:118
Atrazine hydrolysis, hydroxyatrazine metabolite, **151**:122
Atrazine hydroxylation, **151**:122
Atrazine, hysteresis soil sorption, **151**:130
Atrazine, immunoassay detection groundwater, **151**:136
Atrazine K_d (partition coefficient, soil:solution), **151**:126
Atrazine, leaching column, **151**:138, 142
Atrazine, leaching field, **151**:142
Atrazine metabolites, leaching rates, **151**:141
Atrazine metabolites, soil sorption/desorption, **151**:132
Atrazine metabolites, structures, **151**:120
Atrazine, mineralization, **151**:122
Atrazine, movement to groundwater, **151**:136
Atrazine, N-dealkylation, **151**:122
Atrazine, no-observed-effect concentrations soil fauna, **154**:107
Atrazine, offsite movement, **151**:118
Atrazine, percolation to soil depths, **151**:137
Atrazine, persistence application method effects, **151**:152
Atrazine, persistence formulation effects, **151**:151
Atrazine, pH affects soil adsorption, **151**:128
Atrazine, physical/chemical properties, **151**:119
Atrazine, rates of application, **151**:118
Atrazine, risk assessment soil fauna, **154**:91
Atrazine, soil aging, **151**:128
Atrazine, soil binding, **151**:130
Atrazine, soil biodegradation factors, **151**:123
Atrazine, soil desorption, **151**:128

Atrazine, soil fauna field recovery, **154:** 117, 121
Atrazine, soil retention, **151:**125
Atrazine soil sorption, organic carbon influence, **151:**126
Atrazine, soil water content binding, **151:** 131
Atrazine, sorption/desorption soil water content, **151:**132
Atrazine, sorption subsurface soils, **151:** 129
Atrazine, surface runoff leaching, **151:** 150
Atrazine, surface water concentrations, **151:**136
Atrazine, surface water runoff, **151:**134
Atrazine, tile drain field leaching, **151:** 143
Atrazine, tillage effects persistence, **151:** 148
Atrazine, tolerant plant mechanisms, **151:** 119
Atrazine transformation, abiotic, **151:**121
Atrazine transformation, biotic, **151:**122
Atrazine, transformation mechanisms, **151:**119
Atrazine, volatility, **151:**119
Atrazine, volatilization field, **151:**151
Atrazine, volatilization from soil surface, **151:**133
Atrazine, water transport to soil surface, **151:**133
Atrazine, wind-eroded sediments, **151:**133
Australia, DDT/HCH residues foodstuffs, **152:**31
Australia, organochlorine pesticide residues foodstuffs, **152:**30
Automobile tire leachates, aquatic environment effects, **151:**67 ff.
Average daily intake (ADI), organochlorine pesticides various countries, **152:**36
Avian species, Rio Grande Basin, **158:**4
Avian species, scientific names (list), **158:**4
Avian toxicity, polychlorinated alkanes, **158:**104, 106
Azotobacter vinelandii, PHA accumulation, **159:**7

β-HCH, residues animal products India, **152:**14
Bacillus cereus, meat irradiation sensitivity, **154:**24
Bacillus megaterium, PHB synthesis, **159:** 3
Bacillus subtilis mutants, pollutant identification, **159:**63
Bacon, first approved irradiated food, **154:**3
Bacteria, D-Values food irradiation, **154:**7
Bacteria, methylene bromide dehalogenation, **155:**10
Bacteria radiation sensitivity, factors affecting, **154:**8
Bacterial dehalogenation, ^{13}C-NMR spectra, **155:**14
Bacterial pathogens drinking water, **152:** 57 ff.
Bacterial response, food irradiation, **154:**6
BAL (dimercaptopropanol), cadmium chelation therapy, **154:**68
Beach tars, crude oil marine pollution, **153:**101
Benomyl, no-observed-effect concentrations soil fauna, **154:**104
Benomyl, risk assessment soil fauna, **154:** 91, 104
Benomyl, soil faauna field recovery, **154:** 116, 121
Benzene hexachloride, use in Africa, **151:**3
Benzimidazoles, trifluralin degradation products, **153:**10
Benzothiazole, street runoff toxicity, **151:** 106
Benzothiazoles, degradation pathway industrial wastewater, **151:**105
Benzothiazoles, rubber tire antioxidants, **151:**74
Benzothiazoles, rubber tire leachate component, **151:**74, 80, 106
Benzo[α]pyrene, rubber tire fire smoke, **151:**70
Binding, atrazine to soil, **151:**130
Bioaccumulation, cadmium, **154:**57
Bioaccumulation, heavy metals freshwater insect larvae, **158:**129 ff.
Bioaccumulation, polychlorinated alkanes, **158:**88, 91

Bioavailability, heavy metals sediments Gulf of Mexico, **157**:68
Bioconcentration factors, polychlorinated alkanes, **158**:91
Bioconcentration potential, chlorpyrifos, **160**:56
Biodegradable plastics, mixed with starch, **159**:2
Biodegradation, polychlorinated alkanes, **158**:68
Bioindicator, defined, **157**:3
Bioindicators, sediment pollutant assays, **159**:66
BIOLOG system, profiles substrate degradation, **159**:68
Biological degradation, EDTA, **152**:91
Biomagnification factors, polychlorinated alkanes, **158**:93
Bioplastics, degradation properties, **159**:20
Biopolymers, polyhydroxyalkanoates, **159**:2
Biosynthesis, polyhydroxyalkanoates, **159**:7
Biosynthetic pathway, PHB, **159**:10
Biosynthetic pathway, PHBV, **159**:11
Biotransformation, polychlorinated alkanes, **158**:93
Biphasic dissipation, trifluralin soils, **153**:37
Bird eggs, DDT residues US/Europe/Africa compared, **151**:24
Bird eggs, dieldrin residues US/Europe/Africa compared, **151**:25
Bird eggs, organochlorine residues Africa, **151**:15, 24
Birds, DDE residues Rio Grande Basin by year, **158**:11, 12
Birds, DDE studies Rio Grande Basin, **158**:9
Birds, DDT residues US/Europe/Africa compared, **151**:23
Birds, GB toxicity, **156**:142
Birds, mercury residues Rio Grande Basin, **158**:15, 16, 18
Birds, organochlorine residues Africa, **151**:8, 12,
Birds, polychlorinated alkanes toxicity, **158**:104, 106
Birds, Rio Grande Basin, **158**:4
Birds, scientific names (list), **158**:4
Birds, selenium residues Rio Grande Basin, **158**:16, 18
Birds, VX toxicity, **156**:142
Bivalves, modeling metal bioavailability, **151**:39 ff.
Blood agents, chemical warfare, described, **156**:9
Blood ChE inhibition, nerve agent critical toxic effect, **156**:23
Blood ChE inhibition, nerve agent species variation, **156**:21
Blood cholinesterases, nerve agent effects, **156**:20
Blood lead, analytical methods, **159**:30
Blood lead levels, children, **159**:28
Blood lead levels, children vs dogs, **159**:37
Blood lead levels, dogs, **159**:35
Blood lead levels, general population, **159**:32
Blood lead levels, worker exposure, **159**:30
Bootleg spirits, cadmium content, **154**:66
Boron, contamination Rio Grande Basin biota, **158**:10
Bound residues, atrazine soil, **151**:130
Bound residues, ethalfluralin, **153**:70
Bound residues, trifluralin soil, **153**:15
Bound water, defined, **155**:121
Bound-water, natural organic matter relationship, **155**:116
Bound-water organic matter, aquatic ecosystems, **155**:115 ff.
Brain tumors, electromagnetic fields (EMF), **159**:111 ff.
Brazil, major gold mining areas, **157**:27
Brazil, mercury contamination, **157**:25 ff.
Breast cancer, organochlorine associated, **152**:2
Bufencarb, fish kill California, **159**:97
Buffer zones, affect atrazine runoff, **151**:152
Butadiene, occupational inhalation hazard, **151**:82
Butadiene polymers, automobile tire composition, **151**:68

Butadiene rubber, automobile tire composition, **151**:69
Butadiene rubber, chemistry, **151**:69
Butyrylcholinesterase (plasma-ChE), nerve agent effects, **156**:20

Caddisflies (Hydropsychidae), trace metals vs sediment levels, **158**:134
Caddisflies, larval bioaccumulation heavy metals, **158**:131
Caddisflies, Pb, Zn, Cu vs sediment/water levels, **158**:135
Cadmiuim, shark levels Gulf of Mexico, **157**:78
Cadmium, analytical methods, **154**:70, 72
Cadmium, assimilation efficiency marine mussels, **151**:44
Cadmium, bioaccumulation freshwater insect larvae, **158**:132
Cadmium, bioavailability vs chemical species, **154**:65
Cadmium, biological half-life, **154**:63
Cadmium, chelation therapy humans, **154**:68
Cadmium, cigarette smoking intake, **154**:68
Cadmium, contamination Rio Grande Basin biota, **158**:10
Cadmium, EDTA environmental removal, **152**:102
Cadmium, entry human food chain, **154**:66
Cadmium, environmental indicators, **154**:65
Cadmium, environmental monitoring, **154**:65
Cadmium, environmental persistence, **154**:63
Cadmium, farmland soil levels worldwide, **154**:60
Cadmium, farmlands contamination, **154**:55 ff.
Cadmium, high accumulating crops, **154**:62
Cadmium, human dietary exposure, **154**:67
Cadmium, human food/beverage levels, **154**:66
Cadmium, human toxicity/symptoms, **154**:68
Cadmium, industrial environmental sources, **154**:58
Cadmium, industrial uses, **154**:56
Cadmium, inhalation toxicity, **154**:67
Cadmium, interactions other dietary components, **154**:67
Cadmium, irrigation water contamination, **154**:55 ff.
Cadmium, levels in vegetable crops, **154**:64
Cadmium, mutagenic effects, **154**:68
Cadmium, myelin degeneration, **154**:68
Cadmium, natural environmental sources, **154**:57
Cadmium, natural soil levels, **154**:57
Cadmium, natural surface water levels, **154**:59
Cadmium, organism levels Gulf of Mexico, **157**:76
Cadmium, plant absorption from soils, **154**:58
Cadmium, plant levels vs soil pH, **154**:62
Cadmium, pollution Gulf of Mexico, **157**:62
Cadmium, potable water, **154**:61
Cadmium, reduced acetylcholine levels, **154**:68
Cadmium, removal from wastewaters, **154**:60
Cadmium, sediments Gulf of Mexico, **157**:66, 68, 71
Cadmium, sewage sludge/fertilizer regulations, **154**:69
Cadmium, sewage sludge levels, **154**:60
Cadmium, synergistic effect other elements plants, **154**:62
Cadmium, teratogenic effects, **154**:68
Cadmium, toxic effects marine organisms, **157**:84
Cadmium, trace contamination estuaries, **155**:73 ff.
Cadmium, uncontaminated environmental levels, **154**:56
Cadmium, vegetable crop contamination, **154**:55 ff.
Cadmium, water contamination irrigation, **154**:55 ff.

Cadmium, zinc ore relationship, **154**:59
California rivers, chlorpyrifos levels, **160**: 44
Cambodia, organochlorine pesticide residues foodstuffs, **152**:25
Campylobacter, diarrhea leading cause world, **154**:24
Campylobacter jejuni, meat irradiation sensitivity, **154**:24, 34
Carassius auratus, rubber tire leachate toxicity, **151**:77, 88
Carbamate insecticides, toxicity soil arthropods, **154**:123
Carbaryl, no-observed-effect concentrations soil fauna, **154**:103
Carbaryl, risk assessment soil fauna, **154**:91, 102
Carbaryl, soil fauna field recovery, **154**:116, 121
Carbaryl, toxicity to fish, **159**:103
Carbofuran, fish kill California, **159**:97
Carbofuran, no-observed-effect concentrations soil fauna, **154**:101
Carbofuran, risk assessment soil fauna, **154**:91, 100
Carbofuran, soil fauna field recovery, **154**:114, 120
Carbon cycle, polyhydroxyalkanoates, **159**:19
Carboxylesterases (aliesterases), nerve agent effects, **156**:21
Carcinogenicity, CK, **156**:122
Carcinogenicity, GA, **156**:73
Carcinogenicity, GB, **156**:90
Carcinogenicity, HD, **156**:38
Carcinogenicity, HN2, **156**:53
Carcinogenicity, HT, **156**:49
Carcinogenicity, lewisite, **156**:107
Carcinogenicity, polychlorinated alkanes, **158**:114
Carcinogenicity, VX, **156**:61
Caribbean Sea, oil pollution, **153**:91 ff.
CARIPOL marine pollution project, **153**:99
Carp (common), *Cyprinus carpio*, herbicide fish kill, **159**:96
Centropristis striata, rubber tire leachate toxicity, **151**:80, 89

Ceratodon purpureus (moss), heavy metal tolerance, **157**:11
Ceriodaphnia dubia, rubber tire leachate toxicity, **151**:90
Cesium, trace contamination estuaries, **155**:73 ff.
ChE (cholinesterase), nerve agent effects, **156**:20
ChE inhibition, nerve agent potency, **156**:23
Chelation, heavy metals by EDTA/DTPA, **152**:103
Chemical names, ethalfluralin, **153**:67
Chemical names, trifluralin, **153**:5
Chemical nomenclature, ethalfluralin metabolites, **153**:68
Chemical nomenclature, trifluralin metabolites, **153**:63
Chemical structures, conventional petrochemical plastics, **159**:3
Chemical structures, polyhydroxyalkanoates, **159**:4
Chemical warfare agent disposal techniques, **156**:3
Chemical warfare agent non-stockpile sites U.S., **156**:4
Chemical warfare agents, **156**:1 ff.
Chemical warfare agents, acronyms/abbreviations, **156**:152
Chemical warfare agents, air exposure limits, **156**:7
Chemical warfare agents, environmental cleanup, **156**:3
Chemical warfare agents, environmental fate/effects, **156**:127
Chemical warfare agents, estimated reference doses, **156**:8, 146, 150
Chemical warfare agents, glossary, **156**:156
Chemical warfare agents, identifications, **156**:6
Chemical warfare agents, physical/chemical properties, **156**:10, 12, 14, 16
Chemical warfare agents, RfD (reference doses), **156**:8, 146, 150
Chemical warfare agents, storage sites U.S., **156**:4

Chemical warfare blood agents, described, **156:**9
Chemical warfare nerve agents, described, **156:**9
Chemical warfare vesicants, described, **156:**9
Chemical Weapons Convention, **156:**3
Chesapeake Bay, chlorpyrifos water levels, **160:**55
Childhood brain cancer, EMF studies, **159:**114, 115
Childhood brain tumors, appliance use, **159:**124, 126
Childhood brain tumors, descriptive epidemiology, **159:**112
Childhood brain tumors, distance to electrical installations, **159:**122
Childhood brain tumors, electromagnetic fields (EMF), **159:**111 ff.
Childhood brain tumors, five-level wire codes, **159:**120
Childhood brain tumors, magnetic field measurements, **159:**114, 121
Childhood brain tumors, malignant neoplasms, **159:**112
Childhood brain tumors, prenatal appliance exposure, **159:**125
Childhood brain tumors, risk factors, **159:**113
Childhood brain tumors, wire codes HCC/LCC comparison, **159:**119
Childhood cancer, electromagnetic fields, **159:**111
Childhood leukemia, electromagnetic fields, **159:**111
Children, lead blood levels Uruguay, **159:**28
Children, lead exposure sources, **159:**33
China, DDT/HCH MRLs food, **152:**21
China, DDT/HCH residues food, **152:**19
Chinook salmon, *Oncorhynchus tshawytscha*, **159:**99
Chlordane, residues food Australia, **152:**30, 32
Chlordane residues food, country comparisons, **152:**34
Chlorinated diphenyloxides, see also Polychlorinated diphenylethers (PCDEs), **157:**131

Chlorpyrifos, AChE inactivation, **160:**4
Chlorpyrifos, activation to oxon (diag), **160:**13
Chlorpyrifos, additive toxicity other organophosphates, **160:**63
Chlorpyrifos, agricultural application methods, **160:**16
Chlorpyrifos, agricultural pests controlled, **160:**16
Chlorpyrifos, agricultural use by county (illus), **160:**20
Chlorpyrifos, agricultural use patterns, **160:**15
Chlorpyrifos, aquatic ecotoxicology, **160:**59
Chlorpyrifos, bioconcentration factors, **160:**56
Chlorpyrifos, bioconcentration potential, **160:**56
Chlorpyrifos, biotransformation, **160:**57
Chlorpyrifos, California river levels, **160:**44, 98
Chlorpyrifos, Chesapeake Bay water levels, **160:**55
Chlorpyrifos, crops used on, **160:**16
Chlorpyrifos, ecological risk assessment, **160:**1 ff.
Chlorpyrifos, ecological role of sensitive taxa, **160:**107
Chlorpyrifos, ecological significance of effects, **160:**105
Chlorpyrifos, EECs (expected environmental concentrations), **160:**27
Chlorpyrifos, EECs field corn, **160:**29
Chlorpyrifos, effects analysis aquatic systems, **160:**10
Chlorpyrifos, effects on ecological receptors (illus), **160:**4
Chlorpyrifos, environmental concentrations EPA tiers, **160:**27
Chlorpyrifos, environmental persistence, **160:**25
Chlorpyrifos, environmental transformation pathways (diag), **160:**26
Chlorpyrifos, expected environmental concentrations (EECs), **160:**27
Chlorpyrifos, exposure analysis aquatic systems, **160:**8

Chlorpyrifos, exposure characterization aquatic systems, **160**:15
Chlorpyrifos, final acute/chronic calculations, **160**:65, 70, 75
Chlorpyrifos, half-life plant foliage, **160**:25
Chlorpyrifos, half-life soil, **160**:25
Chlorpyrifos, interactive toxicity with other organophosphates, **160**:63
Chlorpyrifos, Lake Erie drainage basin watershed levels, **160**:41, 50, 97, 102
Chlorpyrifos LC_{50}, aquatic ecosystems, **160**:6
Chlorpyrifos, limits of detection water, **160**:38
Chlorpyrifos, littoral enclosure studies, **160**:79
Chlorpyrifos, major crop regions water contamination, **160**:54
Chlorpyrifos, measured concentrations freshwater, **160**:30, 96
Chlorpyrifos, measured concentrations saltwater, **160**:33, 55, 99
Chlorpyrifos, metabolic products (diag), **160**:26
Chlorpyrifos metabolite, trichloro-2-pyridinol (TCP), **160**:7
Chlorpyrifos metabolites, toxicity, **160**:57
Chlorpyrifos, microcosm/mesocosm studies, **160**:78
Chlorpyrifos, NAWQA sampling site levels US (illus), **160**:53
Chlorpyrifos, no-observed-effect concentrations soil fauna, **154**:100
Chlorpyrifos oxon, structure, **160**:13
Chlorpyrifos, physical/chemical properties, **160**:12
Chlorpyrifos, pond studies, **160**:79, 80
Chlorpyrifos, population modeling environmental risk assessment, **160**:110
Chlorpyrifos, probabilistic risk assessment, **160**:84, 96
Chlorpyrifos, reciprocity time vs toxicity fish, **160**:59
Chlorpyrifos, refugia in risk assessment, **160**:92
Chlorpyrifos, return frequency in risk assessment, **160**:92, 109
Chlorpyrifos, risk assessment aquatic environments, **160**:1 ff.
Chlorpyrifos, risk assessment aquatic system process (illus), **160**:8
Chlorpyrifos, risk assessment aquatic systems *a priori*, **160**:8
Chlorpyrifos, risk assessment endpoints, **160**:5
Chlorpyrifos, risk assessment soil fauna, **154**:91, 99
Chlorpyrifos, risk characterization, **160**:82
Chlorpyrifos, San Joaquin River levels (Calif.), **160**:51, 98
Chlorpyrifos, sediment-borne toxicity, **160**:69
Chlorpyrifos, soil fauna field recovery, **154**:114, 120
Chlorpyrifos, spatial/temporal issues in risk assessment, **160**:108
Chlorpyrifos, surface water runoff areas (illus), **160**:23
Chlorpyrifos, synergism with other agrochemicals, **160**:64
Chlorpyrifos, termite control environmental concentrations, **160**:104
Chlorpyrifos, tiered approach environmental concentrations, **160**:27
Chlorpyrifos toxicity, freshwater invertebrate groups, **160**:107
Chlorpyrifos toxicity, freshwater vs saltwater fish, **160**:62
Chlorpyrifos, toxicity pulsed exposures fish, **160**:58
Chlorpyrifos, transport pathways to aquatic systems (illus), **160**:3
Chlorpyrifos, urban use by county (illus), **160**:21
Chlorpyrifos, urban use patterns, **160**:15, 19
Chlorpyrifos, use patterns, **160**:15
Chlorpyrifos, water sampling locations US, **160**:32
Cholinesterase inhibition, nerve agent potency, **156**:23
Chromium (II) model, dehalogenation, **155**:18
Chromium, assimilation efficiency marine mussels, **151**:44

Chromium, bioaccumulation freshwater insect larvae, **158**:133
Chromium, contamination Rio Grande Basin biota, **158**:10
Chromium, organism levels Gulf of Mexico, **157**:76
Chromium, pollution Gulf of Mexico, **157**:65
Chromium, sediments Gulf of Mexico, **157**:66, 68, 71
Chromium, trace contamination estuaries, **155**:73 ff.
Chromosome volume, radiosensitivity correlation, **154**:5
Chromous sulfate, alkyl halide reduction, **155**:21
Chronic toxicity, **156**:32
Chronic toxicity, CK, **156**:116
Chronic toxicity, HN2, **156**:52
Chronic toxicity, HT, **156**:49
Chronic toxicity, T sulfur mustard, **156**:50
Cigarette smoking, cadmium intake, **154**:68
CK (cyanogen chloride), described, **156**:9
CK, acute inhalation toxicity, **156**:114
CK, acute toxicity, **156**:113
CK, carcinogenicity, **156**:122
CK, chronic toxicity, **156**:116
CK, developmental/reproductive effects, **156**:118
CK, ecotoxicology, **156**:145
CK, environmental fate air, **156**:144
CK, environmental fate water, **156**:144
CK, estimated reference dose, **156**:123, 146, 151
CK, fish toxicity, **156**:145
CK, genotoxicity, **156**:122
CK, hydrolytic pathway (diagram), **156**:144
CK, physical/chemical properties, **156**:15
CK, subchronic toxicity, **156**:114
CK, toxicity mechanisms/symptoms, **156**:25
Clay minerals, trace metal adsorption, **155**:85
Clostridium botulinum, D-Values food irradiation, **154**:7, 24

Clostridium botulinum, meat irradiation sensitivity, **154**:24, 38
Clostridium perfringens, meat irradiation sensitivity, **154**:24
Clover (white), atmospheric biomonitor, **157**:6
^{60}Co, food irradiation preservation, **154**:2
Coal, cadmium levels, **154**:58
Coastal metal pollution, Gulf of Mexico, **157**:53 ff.
Cobalt, assimilation efficiency marine mussels, **151**:44
Cobalt, bioaccumulation freshwater insect larvae, **158**:133
Cobalt corrins, alkyl halide reductants, **155**:37
Cobalt, organism levels Gulf of Mexico, **157**:77
Cobalt, pollution Gulf of Mexico, **157**:62
Cobalt, sediments Gulf of Mexico, **157**:69
Cobalt, trace contamination estuaries, **155**:73 ff.
Cobalt-60, food irradiation preservation, **154**:2
Cobinamide-Co(I) complex, Vitamin B$_{12S}$ (illus.), **155**:37
Codex Alimentarius, General Standard for Irradiated Foods, **154**:3
Collembola, dimethoate effects, **154**:95
Collembola, sensitivity to insecticides, **154**:123
Colloidal organic matter, aquatic ecosystems, **155**:115 ff.
Colonization, gut bacteria drinking water disease, **152**:75
Column leaching, atrazine, **151**:138, 142
Comamonas sp., PHA biodegradation, **159**:17
Comamonas testosteroni, PHA biodegradation, **159**:17
Common carp, *Cyprinus carpio*, herbicide fish kill, **159**:96
Coniferous trees, biomonitors atmospheric pollution, **157**:5
Controlled-release formulation, atrazine, **151**:151
Copper, bioaccumulation freshwater insect larvae, **158**:132

Copper, contamination Rio Grande Basin biota, **158**:10
Copper, organism levels Gulf of Mexico, **157**:77
Copper, pollution Gulf of Mexico, **157**:65
Copper, sediments Gulf of Mexico, **157**:69
Copper, trace contamination estuaries, **155**:73 ff.
Coral reefs, marine crude oil spill effects, **153**:109
Cottonwood trees, atmospheric biomonitor, **157**:6
Cr(II) model, dehalogenation, **155**:18
Crassostrea virginica (oysters), heavy metals content Gulf of Mexico, **157**:75
Crassostrea virginica (oysters), heavy metals content Mexican Pacific, **157**:82
Crassostrea virginica, metal bioavailability, **151**:58
Crayfish (*Procambarus clarkii*), cadmium indicator, **154**:65
Crocodile eggs, organochlorine insecticides Africa, **151**:11
Crude oil spills, coral reef effects, **153**:109
Crude oil spills, effects marine organisms, **153**:107
Crude oil spills, Gulf of Mexico, **153**:91 ff.
Crude oil spills, mangrove effects, **153**:108
Crude oil spills, marine direct/indirect effects, **153**:110
Cyanogen chloride, acute toxicity, **156**:113
Cyanogen chloride, described, **156**:9
Cyanogen chloride, environmental fate, **156**:144
Cyanogen chloride, estimated reference dose, **156**:123, 146, 151
Cyanogen chloride, lacrimation, coughing, **156**:25
Cyanogen chloride, physical/chemical properties, **156**:15
Cyanogen chloride, toxicity mechanisms/symptoms, **156**:25

Cyprinodon variegatus, rubber tire leachate toxicity, **151**:86, 89, 94
Cyprinus carpio (carp), molinate herbicide kill, **159**:96
Cytochrome P-450, bromotrichloromethane oxidation, **155**:32
Cytochrome P-450, dehalogenation processes, **155**:16, 31

D-Value (D_{10}), bacteria food irradiation defined, **154**:6
D_{10} (D-Value), bacteria food irradiation defined, **154**:6
D_{10} Values, red meat pathogens, **154**:24
Dairy products, DDT/HCH residues India, **152**:8
Dairy products, irradiation low temperature benefits, **154**:21
Dairy products, irradiation preservation, **154**:21
Daphnia magna, rubber tire leachate toxicity, **151**:83, 90
DDD-p,p', residues farm products India, **152**:14
DDE, bird studies Rio Grande Basin, **158**:9
DDE, bird tissues Rio Grande Basin by year, **158**:11, 12
DDE, eggshell thinning history, **158**:15
DDE, residues amphibians Rio Grande Basin, **158**:20
DDE, residues fish Rio Grande Basin, **158**:21,
DDE, residues invertebrates Rio Grande Basin, **158**:32
DDE, residues mammals Rio Grande Basin, **158**:17, 19
DDE, residues plants Rio Grande Basin, **158**:35
DDE, residues reptiles Rio Grande Basin, **158**:19
DDE, residues sediments Rio Grande Basin, **158**:35, 37
DDE, Rio Bravo Basin pollution, **158**:1 ff.
DDE, Rio Grande Basin pollution, **158**:1 ff.
DDE-p,p', residues farm products India, **152**:14

DDT, ADIs various countries, **152**:36
DDT, air contamination India, **152**:12
DDT, aquatic food chain levels chart Africa, **151**:29
DDT, dehalogenation to DDD, **155**:25
DDT, eggshell thinning, **151**:2
DDT, intake milk products India, **152**:40
DDT, quantity used India, **152**:3
DDT, reproductive failure, **151**:2
DDT, residue reduction foods India, **152**: 6, 17
DDT residues, bird eggs US/Europe/Africa compared, **151**:24
DDT residues, birds Africa, **151**:13, 24
DDT residues, birds US/Europe/Africa compared, **151**:23
DDT residues, crocodile eggs Africa, **151**:11
DDT, residues dairy products India, **152**:8
DDT, residues drinking water India, **152**:12
DDT residues, fish Africa, **151**:18, 29
DDT residues, fish US/Europe, **151**:20
DDT residues, fish US/Europe/Africa compared, **151**:26
DDT, residues food China, **152**:19
DDT, residues food grains India, **152**:3
DDT, residues food India, **152**:4
DDT, residues food Thailand, **152**:22
DDT residues, mammals Africa, **151**:10
DDT, residues meat products India, **152**:11
DDT, residues spices/beverages India, **152**:7
DDT, residues vegetables/fruits India, **152**:6
DDT-p,p', residues farm products India, **152**:14
DDT-R (total DDT residues), **151**:6
Dealkylation, atrazine, **151**:122
Dealkylation, trifluralin degradation, **153**:14
Deamination, atrazine, **151**:122
Deethylatrazine, atrazine metabolite, **151**:120
Deethyldeisopropylatrazine, atrazine metabolite, **151**:120
Deethylhydroxyatrazine, atrazine metabolite, **151**:120

Degradation pathway, atrazine, **151**:121
Dehalogenation, abiotic processes, **155**:5
Dehalogenation, biotic processes, **155**:10
Dehalogenation, by hemes, **155**:23
Dehalogenation, by iron(II) porphyrins, **155**:22
Dehalogenation, chemistry/mechanism environmental, **155**:1 ff.
Dehalogenation, Cr(II) model, **155**:18
Dehalogenation, environmental, **155**:1 ff., 5
Dehalogenation, hydrolysis/substitution environmental, **155**:5
Dehalogenation, naturally occurring, **155**:1 ff.
Dehalogenation, oxidation environmental, **155**:9
Dehalogenation, photohydrolysis environmental, **155**:6
Dehalogenation, reduction environmental, **155**:7
Dehalogenation, site reactivity probes, **155**:49
Dehydrogenase activity assays, artificial electron acceptors, **159**:60
Dehydrogenases, use in toxicological studies, **159**:59
Deisopropylatrazine, atrazine metabolite, **151**:120
Delac MOR, major rubber product breakdown, **151**:74
Delayed neuropathy, nerve agents, **156**:19
Delayed toxicity, HD, **156**:34
Delayed toxicity, HN2, **156**:52
Dephosphorylation, chlorpyrifos AChE, **160**:13
Deuteroheme, dehalogenation oxidation rates, **155**:36
Developmental/reproductive effects, CK, **156**:118
Developmental/reproductive effects, GA, **156**:73
Developmental/reproductive effects, GB, **156**:88
Developmental/reproductive effects, HD, **156**:35
Developmental/reproductive effects, HN2, **156**:52
Developmental/reproductive effects, lewisite, **156**:105, 109

Developmental/reproductive effects, VX, **156**:60
Diarrhea, *Campylobacter* leading world cause, **154**:24
Diarrhea causing bacteria, *Aeromonas*, **152**:63
Dibromochloropropane, photohydrolysis, **155**:7
Dibromoethane, photohydrolysis environmental, **155**:8
Dieldrin residues, bird eggs US/Europe/Africa compared, **151**:25
Dieldrin residues, crocodile eggs Africa, **151**:11
Dieldrin residues, fish Africa, **151**:18, 29
Dieldrin residues, fish US/Europe, **151**:20
Dieldrin residues, fish US/Europe/Africa compared, **151**:26
Dieldrin, residues food India, **152**:15
Dieldrin, residues food Thailand, **152**:22
Dieldrin residues, mammals Africa, **151**:10
Dietary intakes, organochlorine pesticides Australia, **152**:42
Dietary intakes, organochlorine pesticides India, **152**:38
Dietary intakes, organochlorine pesticides Thailand, **152**:41
Diethylenetriaminepentaacetic acid (DTPA), environmental fate, **152**:85 ff.
Dilatometry, bound water measuring method, **155**:121
Dimethoate, no-observed-effect concentrations soil fauna, **154**:96
Dimethoate, risk assessment soil fauna, **154**:91, 95
Dimethoate, soil fauna field recovery, **154**:112, 120
Dinitroaniline herbicides, ethalfluralin, **153**:65 ff.
Dinitroaniline herbicides, trifluralin, **153**:1 ff.
Diphenylamine, new rubber tire leachate, **151**:78
Dissolved metal uptake rate constants (K_u), marine mussels, **151**:48
Dissolved organic carbon, affects metal bioavailability aquatic organisms, **151**:40
Dissolved organic carbon, atrazine binding soil, **151**:131
Dissolved organic carbon, salinity function estuaries, **155**:87
Dissolved organic matter, aquatic ecosystems, **155**:115 ff.
Dissolved petroleum hydrocarbons, marine Gulf of Mexico, **153**:100
Distance, EMF child brain tumors, **159**:120, 122
Distance to electrical installations, child brain tumors, **159**:122
Distilled mustard (Agent HD), **156**:11
Dogs, blood lead levels Uruguay, **159**:35
Dose-response modeling, drinking water pathogens, **152**:72
Drinking water, heavy metals permissible levels US, **157**:85
Drinking water, opportunistic bacterial pathogens, **152**:57 ff.
Drinking water pathogens, infectious dose levels, **152**:73
Drinking water pathogens, ranking of importance, **152**:71
DTPA (diethylenetriaminepentaacetic acid), cadmium chelation therapy, **154**:68
DTPA (diethylenetriaminepentaacetic acid), environmental fate, **152**:85 ff.
DTPA, biological degradation, **152**:92
DTPA, commercial uses, **152**:85
DTPA, effect heavy metal toxicity, **152**:101
DTPA, effects aquatic organisms, **152**:97
DTPA, environmental desorption heavy metals, **152**:100
DTPA, molecular structure, **152**:86
DTPA, release to aquatic environments, **152**:85
Dumpsite soil, alkyl halide probe responses, **155**:56
Dysprosium, trace contamination estuaries, **155**:73 ff.

Earthworm burrows, atrazine water channels, **151**:137
Earthworms, carbofuran effects, **154**:101
Earthworms, chlorpyrifos effects, **154**:100
Earthworms, dimethoate effects, **154**:95

Earthworms, parathion-ethyl effects, **154**: 98
Earthworms, sensitivity organophosphate insecticides, **154**:123
Ecdyonuridae (Mayflies), larval bioaccumulation heavy metals, **158**:132
Ecdyonuridae (Mayflies), trace metals vs sediment levels, **158**:134
Ecological risks, EDTA/DTPA, **152**:94
Ecotoxicological effects, EDTA, **152**:95
Ecotoxicological recovery time, pesticides soil fauna, **154**:89
Ecotoxicological risk assessment, pesticides soil fauna, **154**:83 ff.
Ecotoxicology, CK, **156**:145
Ecotoxicology, lewisite, **156**:143
Ecotoxicology, mustard agents, **156**:132
Ecotoxicology, nerve agents, **156**:141
EDTA (ethylenediaminetetraacetic acid), cadmium chelation therapy, **154**:68
EDTA (ethylenediaminetetraacetic acid), environmental fate, **152**:85 ff.
EDTA, analytical methods, **152**:86
EDTA, bactericide mode of action, **152**: 98
EDTA, banned from phosphate-free detergents, **152**:86
EDTA, behavior during water treatment, **152**:86
EDTA, biological degradation, **152**:91
EDTA, chemical degradation, **152**:93
EDTA, commercial uses, **152**:85
EDTA, degradation by microorganisms, **152**:88
EDTA, desorption environmental heavy metals, **152**:100
EDTA, ecological risks, **152**:94
EDTA, ecotoxicological effects, **152**:95
EDTA, effect heavy metal toxicity, **152**: 101
EDTA, effects aquatic organisms, **152**:96
EDTA, environmental heavy metals desorption, **152**:100
EDTA, environmental occurrence, **152**:89
EDTA, eutrophic effects, **152**:95
EDTA, heavy metal chelation, **152**:103
EDTA, heavy metals toxicity effect, **152**: 101
EDTA, induces environmental Zn deficiency, **152**:98

EDTA, inhibits sulfide metal precipitation, **152**:87
EDTA, molecular structure, **152**:86
EDTA, nitrogen eutrophic effects, **152**:95
EDTA, nondegradable quality, **152**:87
EDTA, occurrence aquatic biota, **152**:89
EDTA, occurrence natural waters, **152**:89
EDTA, ozonation removal from water, **152**:88
EDTA, ozonation sensitivity, **152**:88
EDTA, photochemical degradation, **152**: 93
EDTA, release to aquatic environment, **152**:85
EDTA, removal in waste water treatment, **152**:87
EDTA, secondary stream pollution, **152**: 100
EDTA-metal complexes, biological degradation, **152**:92
EEC, expected environmental concentration chlorpyrifos, **160**:27
Efflux, metals marine mussels, **151**:50
Eichhornia crassipes (water hyacinth), mercury biomonitor, **157**:34
Eisenia fetida, lindane effects, **154**:94
Electrical appliances, child brain tumors, **159**:124, 126
Electromagnetic fields (EMF), child brain tumors, **159**:111 ff.
Electron accelerator, grain irradiation, **154**:16
Electron acceptors, dehydrogenase activity assays, **159**:60
Electron transport system activity, toxicological studies, **159**:59
Embryo toxicity, polychlorinated alkanes, **158**:105
EMF, child brain cancer studies, **159**:114
EMF, child brain tumor exposure assessment, **159**:118
EMF, child cancer, **159**:111
EMF, child leukemia, **159**:111
EMF, distance child brain tumors, **159**: 120, 122
EMF epidemiological literature, **159**:112
EMF, see Electromagnetic fields, **159**:111
EMF, wire codes child brain tumors, **159**: 118
Endosulfan residues, fish Africa, **151**:19

Endrin residues, fish Africa, **151**:19
Environmental cadmium, uncontaminated levels, **154**:56
Environmental dehalogenation, **155**:5
Environmental fate modeling, polychlorinated alkanes, **158**:115
Environmental fate, mustard agents, **156**:127
Environmental lead, Uruguay studies, **159**:28
Enzyme induction, PCDEs, **157**:139
Enzyme induction, polychlorinated alkanes, **158**:93, 95
Ephemeridae (Mayflies), larval bioaccumulation heavy metals, **158**:132
Ephemeridae (Mayflies), trace metals vs sediment levels, **158**:134
EqP (equilibrium partitioning), metals marine mussels, **151**:41
Equilibrium partitioning (EqP), metals marine mussels, **151**:41
Equilibrium Partitioning Toxicity Tests, USEPA, **151**:75
Erbium, trace contamination estuaries, **155**:73 ff.
Escherichia coli (recombinant), PHA accumulation, **159**:7, 14
Escherichia coli (recombinant), PHA production, **159**:14
Escherichia coli 0157:H7, meat patty radiation sensitivity, **154**:24
Escherichia coli, D-Values food irradiation, **154**:7, 24
Escherichia coli, radiation/temperature sensitivity, **154**:10
Escherichia coli, TOXI-Chromotest® tire leachate, **151**:85, 99
Estimated reference dose, CK (cyanogen chloride), **156**:123, 146, 151
Estimated reference dose, GA (tabun), **156**:74, 146, 150
Estimated reference dose, GB (sarin), **156**:91, 146, 150
Estimated reference dose, GD (soman), **156**:100, 146, 150
Estimated reference dose, HD, **156**:45, 146, 149
Estimated reference dose, HN2, **156**:53, 146

Estimated reference dose, HT, **156**:50, 146
Estimated reference dose, lewisite, **156**:108, 146, 151
Estimated reference dose, VX, **156**:62, 146, 149
Estimated reference doses, chemical warfare agents, **156**:146
Estuaries, principal trace metal pollution sources, **155**:74
Estuaries, trace metal bottom sediment levels, **155**:80
Estuaries, trace metal chemical forms, **155**:76
Estuaries, trace metal-sediment pollution, **155**:73 ff.
Estuarine bottom sediments, trace metal contaminants, **155**:92
Estuarine bottom sediments, trace metal profiles, **155**:95
Estuarine bottom sediments, trace metal removal processes, **155**:96
Estuarine fish, rubber tire leachate toxicity, **151**:96
Estuarine organisms, rubber tire leachate toxicity, **151**:86
Estuarine trace metal pollutants, listed, **155**:80
Estuarine trace metal pollution assessment, **155**:73 ff.
Ethalfluralin, ^{14}C-labeled field studies, **153**:74
Ethalfluralin, aerobic soil metabolism, **153**:69
Ethalfluralin, air photolysis, **153**:68
Ethalfluralin, anaerobic soil metabolism, **153**:69
Ethalfluralin, aquatic metabolism, **153**:69
Ethalfluralin, biological availability, **153**:81
Ethalfluralin, biological transformation, **153**:69 ff.
Ethalfluralin, chemical name, **153**:67
Ethalfluralin, chemical structure, **153**:88
Ethalfluralin, chemically reactive transport modeling, **153**:85
Ethalfluralin, crops annual usage, **153**:66
Ethalfluralin dissipation pathway chart, **153**:71

Ethalfluralin, environmental dissipation, **153**:73 ff.
Ethalfluralin, environmental fate, **153**:65 ff.
Ethalfluralin, environmental mobility, **153**:72 ff.
Ethalfluralin, Fruendlish coefficients soils, **153**:72
Ethalfluralin, fugacity modeling, **153**:85
Ethalfluralin, half-life field studies, **153**:80
Ethalfluralin, Henry's Law Constant, **153**:67
Ethalfluralin, hydrolysis, **153**:67
Ethalfluralin, metabolites, **153**:70
Ethalfluralin metabolites, chemical nomenclature, **153**:63
Ethalfluralin, mode of action, **153**:66
Ethalfluralin, photodecomposition, **153**:67
Ethalfluralin, photolysis, **153**:67
Ethalfluralin, physicochemical properties, **153**:67
Ethalfluralin, soil adsorption/desorption, **153**:72
Ethalfluralin, soil bound residues, **153**:70
Ethalfluralin, soil leaching, **153**:72
Ethalfluralin, soil photolysis, **153**:67
Ethalfluralin, soil residue decline, **153**:77
Ethalfluralin, soil transformation pathway chart, **153**:71
Ethalfluralin, vegetation residues, **153**:82
Ethalfluralin, volatile flux field applied, **153**:77
Ethalfluralin, volatilization from soil, **153**:73
Ethalfluralin, water photolysis, **153**:67
Ethylene dibromide, photohydrolysis environmental, **155**:8
Ethylenediaminetetraacetic acid (EDTA), environmental fate, **152**:85 ff.
Eurasian otter (*Lutra lutra*), PCBs effects, **157**:95 ff.
Eurasian otter, population declines Europe, **157**:95
Eurasian otter, severe decline cause Sweden, **157**:112
Eurasian otter vs American River otter, PCB levels, **157**:105

Europium, trace contamination estuaries, **155**:73 ff.
Eutrophic effects, EDTA/DTPA, **152**:95
Evergreen plants, preferred atmospheric biomonitors, **157**:7
EXAMS, model described, **160**:115
Expected environmental concentrations (EECs), **160**:27
Exposure assessment, EMF child brain tumors, **159**:118
Exposure limits (air), chemical warfare agents, **156**:7

F-430 factor, alkyl halide reactions, **155**:38
Factor F-430, alkyl halide reactions, **155**:38
FAO/WHO MRLs, organochlorines foods, **152**:18
Fathead minnow, *Pimephales promales*, **159**:99
Fathead minnow, rubber tire leachate toxicity, **151**:83, 87
Fathead minnows, chlorpyrifos toxicity, **160**:60
Fell protein oxidation, alkyl halides, **155**:29
Fentin, no-observed-effect concentrations soil fauna, **154**:108
Fentin, risk assessment soil fauna, **154**:91
Ferret (*Mustela putorius furo*), PCB sensitivity, **157**:108
Fetal effects, HD, **156**:36
Field capacity, soil water atrazine binding, **151**:131
Field leaching, atrazine, **151**:142
Filterable particulate organic matter (FPOM), (illus.), **155**:118
Final acute value, chlorpyrifos, **160**:66, 70
Final chronic value, chlorpyrifos, **160**:69, 75
Fish, CK toxicity, **156**:145
Fish consumption, mercury hair concentrations, **157**:43
Fish, DDE residues Rio Grande Basin, **158**:21
Fish, estuarine rubber tire leachate toxicity, **151**:96

Fish, freshwater rubber tire leachate toxicity, **151**:83
Fish, GA toxicity, **156**:141
Fish, GB toxicity, **156**:141
Fish genera, organochlorine residues Africa, **151**:18
Fish, heavy metal levels Mexican Pacific, **157**:82
Fish, human methylmercury poisoning, **157**:37
Fish, marine rubber tire leachate toxicity, **151**:97
Fish, mercury biomonitors, **157**:36
Fish, mercury contamination levels Brazil, **157**:38
Fish, mercury residues Rio Grande Basin, **158**:27
Fish, mustard agents toxicity, **156**:132
Fish, nerve agent toxicity, **156**:141
Fish, organochlorine residues Africa, **151**:18, 26
Fish, polychlorinated alkanes toxicity, **158**:100
Fish, Rio Grande Basin, **158**:5
Fish, rubber tire leachate toxicity, **151**:77, 83
Fish, scientific names (list), **158**:5
Fish, selenium residues Rio Grande Basin, **158**:30
Fish, VX toxicity, **156**:141
Floating tars, crude oil marine pollution, **153**:101
Flocculation, trace metal movement rivers, **155**:83
Fluorocarbons, volcanic synthesis, **155**:3
Fluorochlorocarbons, volcanic synthesis, **155**:3
Fluorophosphonate chemical warfare agents, **156**:137
Folsomia candida, lindane effects, **154**:94
Food additive, irradiation classified as, **154**:3
Food irradiation, bacterial response, **154**:6
Food irradiation, cell concentration effects, **154**:10
Food irradiation, costs, **154**:14
Food irradiation, dairy products, **154**:21
Food irradiation, dose application effects, **154**:11

Food irradiation, dose rate effects, **154**:10
Food irradiation, doses cause fruit damage vs rot inhibition, **154**:13
Food irradiation, fruits and vegetables, **154**:12
Food irradiation, grains, **154**:15
Food irradiation, oxygen atmosphere effects, **154**:9
Food irradiation, pH effects, **154**:9
Food irradiation, poultry meat, **154**:32
Food irradiation, red meats, **154**:22
Food irradiation, seafood preservation, **154**:35
Food irradiation, shelf-life extension, **154**:14
Food irradiation, spices, **154**:18
Food irradiation, temperature effects, **154**:9
Food preservation, ionizing radiation, **154**:1 ff.
Food safety, irradiation health risks, **154**:3
Food spoilage microorganisms, irradiation control, **154**:19
Foodborne pathogens red meat, D_{10} irradiation values, **154**:24
Free water, defined, **155**:120
Free water, natural organic matter relationship, **155**:116
Freshwater fish, rubber tire leachate toxicity, **151**:83
Freshwater insect larvae, heavy metal bioaccumulation, **158**:129 ff.
Freshwater invertebrates, rubber tire leachate toxicity, **151**:97
Freundlich isotherms clay adsorption, atrazine, **151**:127
Fruendlich coefficients (K_f), ethalfluralin soils, **153**:72
Fruit irradiation, ripening inhibition, **154**:14
Fruit irradiation, senescence delay, **154**:2
Fruit, maximum tolerable preservation irradiation, **154**:13
Fruit, ripening inhibition irradiation, **154**:14
Fruit storage rots, irradiation dose required for control, **154**:13
Fruits/vegetables, irradiation preservation, **154**:12
Fugacity modeling, ethalfluralin, **153**:85

GA (tabun), acute toxicity, **156**:68
GA (tabun), carcinogenicity, **156**:73
GA (tabun), described, **156**:9
GA (tabun), developmental/reproductive effects, **156**:73
GA (tabun), estimated reference dose, **156**:74, 146, 150
GA (tabun), genotoxicity, **156**:74
GA (tabun), LD$_{50}$ values, **156**:68
GA (tabun), neurotoxicity, **156**:72
GA (tabun), physical/chemical properties, **156**:13, 14
GA (tabun), plasma-ChE subchronic activity, **156**:71
GA (tabun), RBC-ChE subchronic activity, **156**:70
GA (tabun), subchronic toxicity, **156**:68
GA, fish toxicity, **156**:141
GA, hydrolysis, **156**:136
GA, hydrolytic pathway (diagram), **156**:136
GA, hydrolytic products, **156**:137
Gadolinium, trace contamination estuaries, **155**:73 ff.
Gallium, trace contamination estuaries, **155**:73 ff.
Gauss (G), measure of magnetic field intensity, **159**:116
GB (sarin), acute toxicity, **156**:78
GB (sarin), carcinogenicity, **156**:90
GB (sarin), chronic toxicity, **156**:85
GB (sarin), described, **156**:9
GB (sarin), developmental/reproductive effects, **156**:88
GB (sarin), estimated reference dose, **156**:91, 146, 150
GB (sarin), genotoxicity, **156**:90
GB (sarin), LD$_{50}$ values, **156**:79
GB (sarin), neurotoxicity, **156**:87
GB (sarin), physical/chemical properties, **156**:13
GB (sarin), rat tracheitis incidence, **156**:86
GB (sarin), subchronic toxicity, **156**:80
GB (sarin) type I, RBC-ChE subchronic activity, **156**:83
GB (sarin) type II, plasma ChE subchronic activity, **156**:84
GB (sarin) type II, RBC-ChE subchronic activity, **156**:82
GB (serin), estimated reference dose, **156**:146, 150
GB, bird toxicity, **156**:142
GB, fish toxicity, **156**:141
GB, hydrolysis, **156**:136
GB, hydrolytic pathway (diagram), **156**:137
GB, hydrolytic products, **156**:137
GD (soman), acute toxicity, **156**:94
GD (soman), estimated reference dose, **156**:100, 146, 150
GD (soman), genotoxicity, **156**:100
GD (soman), LD$_{50}$ values, **156**:94
GD (soman), neurotoxicity, **156**:98
GD (soman), plasma-ChE subchronic activity, **156**:97
GD (soman), RBC-ChE subchronic activity, **156**:96
GD (soman), subchronic toxicity, **156**:95
GD, hydrolytic pathway (diagram), **156**:138
GD, hydrolytic products, **156**:137
GD, photolysis, **156**:134
GENEEC, EPA model described, **160**:115
GENEEC, pesticide transport model, **153**:86
General Standard for Irradiated Foods, Codex Alimentarius, **154**:3
Genotoxicity, CK, **156**:122
Genotoxicity, GA, **156**:74
Genotoxicity, GB, **156**:90
Genotoxicity, GD, **156**:100
Genotoxicity, HD, **156**:43
Genotoxicity, HT, **156**:50
Genotoxicity, lewisite, **156**:108
Genotoxicity, T sulfur mustard, **156**:51
Genotoxicity testing, pollutant assay, **159**:62
Genotoxicity tests, rubber tire leachate, **151**:92
Genotoxicity, VX, **156**:61
Germanium, trace contamination estuaries, **155**:73 ff.
GLEAMS, model described, **160**:115
Global element cycles, sediment pollutants impact, **159**:65
Glossary, chemical warfare agents, **156**:156
Glutathion-atrazine, soil sorption, **151**:132

Gold mining areas, major in Brazil, **157:** 27
Gold mining, environmental mercury contamination, **157:**28
Goldfish, rubber tire leachate toxicity, **151:**77, 88
Goldsmiths, mercury exposure, **157:**44
Grains, irradiation insect control, **154:**15
Grains, irradiation nutrient stability, **154:** 17
Grains, irradiation preservation, **154:**15
Grains, irradiation rancidity reduction, **154:**18
Gram-negative bacteria, radiation sensitivity, **154:**35
Gram-positive bacteria, radiation sensisitivity, **154:**35
Grass, biomonitors atmospheric pollution, **157:**6
Gray (gy), unit of absorbed radiation, **154:**2
Groundwater, atrazine contamination, **151:**118
Groundwater, atrazine movement, **151:** 136
Gulf of Mexico, coastal metal pollution, **157:**53 ff.
Gulf of Mexico, heavy metal pollution, **157:**61, 64
Gulf of Mexico, heavy metal pollution sources, **157:**55
Gulf of Mexico, heavy metal rivers discharging, **157:**58
Gulf of Mexico, oil pollution, **153:**91 ff.
Guppies, rubber tire leachate toxicity, **151:**77, 88
Gut colonization, relationship drinking water disease, **152:**75
Gy (gray), unit of absorbed radiation, **154:**2

H sulfur mustard, described, **156:**9
H_2O_2 (hydrogen peroxide) production, food irradiation, **154:**4
Hair, human methylmercury exposure, **157:**41
Half-lives, ethalfluralin field studies, **153:** 80
Half-lives, trifluralin soil, **153:**36

Halocarbons, atmospheric oxidation, **155:** 9
Halocarbons, synthesis in sea, **155:**2
Halogen cycle, **155:**2
Halogen cycle, environmental destruction, **155:**5
Halogen cycle, environmental synthetic leg, **155:**4
Halogenated compounds, sea derived, **155:**2
Halogenation, alpha/beta, **155:**19
Haloiron(III) porophyrin, **155:**25
Haloorganics, abundance in sea, **155:**2
Haloorganics, environmental dehalogenation, **155:**1 ff.
Haloorganics, sea synthesis, **155:**2
Haloorganics, volcanic synthesis, **155:**3
Hamburger patties, irradiation shelf-life extension, **154:**26
Hazard quotients, chlorpyrifos risk characterization, **160:**83, 93
HCB (hexachlorobenzene), residues food Australia, **152:**33
HCB, ADIs various countries, **152:**36
HCC, see High current configuration, **159:**114
HCH-α, β, residues farm products India, **152:**14
HCH (hexachlorocyclohexane), use in Africa, **151:**3
HCH, ADIs various countries, **152:**36
HCH, air contamination India, **152:**12
HCH, quantity used India, **152:**3
HCH, residue reduction food India, **152:** 6, 17
HCH residues, crocodile eggs Africa, **151:**11
HCH, residues dairy products India, **152:** 8
HCH, residues drinking water India, **152:** 12
HCH residues, fish Africa, **151:**19
HCH residues, fish US/Europe, **151:**20
HCH, residues food China, **152:**19
HCH, residues food grains India, **152:**3
HCH, residues food India, **152:**4
HCH, residues food Thailand, **152:**22
HCH, residues meat products India, **152:** 11

HCH, residues spices/beverages India, **152**:7
HCH, residues vegetables/fruits India, **152**:6
HD, acute toxicity, **156**:30
HD, carcinogenicity, **156**:38
HD, chronic toxicity, **156**:32
HD, delayed toxicity, **156**:34
HD, developmental/reproductive effects, **156**:35
HD, described (sulfur mustard), **156**:9, 11
HD, distilled mustard, **156**:11
HD, environmental fate air, **156**:128
HD, environmental fate soil, **156**:131
HD, environmental fate water, **156**:128
HD, estimated reference dose, **156**:45, 146, 149
HD, fetal effects, **156**:36
HD, genotoxicity, **156**:43
HD, human carcinogenicity, **156**:38
HD, hydrolytic products, **156**:130
HD, physical/chemical properties, **156**:11
HD, primary hydrolytic pathway (diagram), **156**:129
HD, skin tumors, **156**:43
HD, subchronic toxicity, **156**:31
HD sulfur mustard, described, **156**:9
HD sulfur mustard, physical/chemical properties, **156**:11
HD, teratogenicity, **156**:46
HD, toxicity effects dogs, **156**:34
HD, vapor effects eyes, **156**:33
HDPE, chemical structure, **159**:3
HDPE, see High density polyethylene, **159**:3
Heat-shock proteins, sediment pollutant assays, **159**:66
Heavy metal bioaccumulation, freshwater insect larvae, **158**:129 ff.
Heavy metal bioaccumulation methodology, insect larvae, **158**:130
Heavy metal desorption, EDTA, **152**:100
Heavy metals, bioavailability sediments Gulf of Mexico, **157**:68
Heavy metals, contamination Rio Grande Basin biota, **158**:10
Heavy metals, discharging rivers to Gulf of Mexico, **157**:58

Heavy metals, effects in aquatic environment (illus.), **157**:57
Heavy metals, leaching from rubber tires, **151**:76
Heavy metals, levels Gulf of Mexico, **157**:64
Heavy metals, organism levels Mexican Pacific coast, **157**:82
Heavy metals, permissible levels US drinking water, **157**:85
Heavy metals, sediment levels Gulf of Mexico, **157**:67
Heavy metals, sediment levels vs freshwater insect larvae, **158**:134
Heavy metals, sediments continental shelf Gulf of Mexico, **157**:71
Heavy metals, sediments Mexican Pacific coast, **157**:78, 80
Heavy metals, toxic effects marine organisms, **157**:81
Hemeprotein conformations (illus.), **155**:28
Hemeproteins, dehalogenation processes, **155**:27
Hemes, alkyl halide reduction, **155**:23
Henry's Law Constant, atrazine, **151**:133
Henry's Law Constants, polychlorinated alkanes, **158**:64
Heptachlor/epoxide, ADIs various countries, **152**:36
Heptachlor, residues food India, **152**:15
Herbicide effects, Sacramento River & Delta, **159**:95 ff.
Herbicides, chemical names (list), **158**:7
Herbicides, direct effects soil fauna, **154**:108
Herbicides, effects aquatic ecosystems, **159**:95 ff.
Herbicides, ethalfluralin environmental fate, **153**:65 ff.
Herbicides, trifluralin environmental fate, **153**:1 ff.
Heterotrophic plate count (HPC) bacteria, tap water, **152**:58
Heterotrophic plate count (HPC) bacteria, water public health risk, **152**:57 ff.
Hexachlorobenzene (HCB), residues food Australia, **152**:33
Hexachlorocyclohexane (HCH), use in Africa, **151**:3

High current configuration (HCC), based on wire configuration code, **159**:114
High-density polyethylene (HDPE), chemical structure, **159**:3
Hillsborough Bay, Florida, sediment trace metal study, **155**:100
HN2 (nitrogen mustard), physical/chemical properties, **156**:12
HN2, acute toxicity, **156**:51
HN2, carcinogenicity, **156**:53
HN2, chronic toxicity, **156**:52
HN2, delayed toxicity, **156**:52
HN2, developmental/reproductive effects, **156**:52
HN2, estimated reference dose, **156**:53, 146
HN2 nitrogen mustard, described, **156**:9
Hospital-acquired (nosocomial) pathogens, **152**:59
Household electrical appliances, child brain tumors, **159**:124, 126
HPC (heterotrophic plate count) bacteria, tap water, **152**:58
HPC (heterotrophic plate count) bacteria, water public health risk, **152**:57 ff.
HPC agar bacterial culture medium, **152**:58
HPC bacteria, oral infectious doses humans, **152**:76
HPLC analysis, plant atmospheric biomonitors, **157**:8
HT, acute toxicity, **156**:49
HT, carcinogenicity, **156**:49
HT, chronic toxicity, **156**:49
HT, described (sulfur mustard), **156**:9, 11
HT, developmental/reproductive effects, **156**:49
HT, estimated reference dose, **156**:50, 146
HT, genotoxicity, **156**:50
HT, physical/chemical properties, **156**:11
HT sulfur mustard, described, **156**:9
HT sulfur mustard, physical/chemical properties, **156**:11
Human carcinogenicity, HD, **156**:38
Human diets, cadmium intake, **154**:67
Human exposure to lead, Uruguay, **159**:25 ff.
Human exposure to mercury, Amazon, **157**:45

Humified organic matter, atrazine bound residues, **151**:130
Hydrogen peroxide, formed during food irradiation, **154**:4
Hydrolysis, atrazine, **151**:122
Hydrolysis, environmental dehalogenation processes, **155**:5
Hydrolysis, polychlorinated alkanes, **158**:68
Hydrolytic pathway, GA (diagram), **156**:136
Hydrolytic pathway, GB (diagram), **156**:137
Hydrolytic pathway, GD (diagram), **156**:138
Hydrolytic pathway, HD (diagram), **156**:129
Hydrolytic pathway, VX (diagram), **156**:135
Hydrolytic products, GA, GB, GD, **156**:137
Hydrolytic products, HD, **156**:130
Hydrolytic products, VX, **156**:135
Hydroperoxyl radical, formed during food irradiation, **154**:5
Hydropsychidae (Caddisflies), trace metals vs sediment levels, **158**:134
Hydrous metal oxides, trace metal adsorption, **155**:86
Hydroxyatrazine, atrazine metabolite, **151**:120
Hydroxybutyric acid, chemical structure, **159**:5
Hydroxyl radical, formed during food irradiation, **154**:5
Hydroxylation, atrazine, **151**:122
Hydroxyvaleric acid, chemical structure, **159**:5
Hylocomium splendens (moss), atmospheric biomonitor, **157**:4, 13
Hypnum cuppressiforme (moss), atmospheric biomonitor, **157**:4, 13
Hysteresis, atrazine soil sorption, **151**:130

Ilyobacter delafieldii, PHA biodegradation, **159**:17
Immunoassay, atrazine detection groundwater, **151**:136
Immunosupression, PCDEs, **157**:139

Incineration, plastics waste energy utilization, **159**:2
India, organochlorine food residue limits, **152**:17
Indicator gardens, atmospheric biomonitors, **157**:6
Indonesia, DDT/HCH residues food, **152**:24
Indonesia, organochlorine pesticide residues food, **152**:24
Inhalation toxicity, cadmium, **154**:67
Insect disinfestation, irradiated wheat, **154**:3
Insect larvae, heavy metal bioaccumulation, **158**:129 ff.
Insect resistance, irradiation, **154**:16
Insecticide residues, organochlorine African fauna, **151**:1 ff.
Internet, atrazine biodegradation pathway, **151**:122
Invertebrates (freshwater), chlorpyrifos toxicity, **160**:107
Invertebrates, DDE residues Rio Grande Basin, **158**:32
Invertebrates, freshwater rubber tire leachate toxicity, **151**:97
Invertebrates, mercury residues Rio Grande Basin, **158**:34
Invertebrates, Rio Grande Basin, **158**:6
Invertebrates, scientific names (list), **158**:6
Invertebrates, selenium residues Rio Grande Basin, **158**:34
Invertebrates, trace element residues Rio Grande Basin, **158**:33
Iodine, trace contamination estuaries, **155**:73 ff.
Ionizing radiation, biological effects, food, **154**:5
Ionizing radiation, food preservation, **154**:1 ff.
Ionizing radiation, food preservation sources, **154**:2
Ionizing radiation, insect resistance, **154**:16
Ionizing radiation, military field rations, **154**:3
Iron, bioaccumulation freshwater insect larvae, **158**:132
Iron oxides, trace metal adsorption estuaries, **155**:88

Iron porphyrins, environmental dehalogenation processes, **155**:22
Iron pyrite, role in environmental dehalogenation, **155**:7
Iron, solution uptake marine mussels, **151**:48
Iron, trace contamination estuaries, **155**:73 ff.
Iron(II) deuteroporphyrin, alkyl halide reduction, **155**:26
Iron(II) porphyrins, alkyl halide oxidation rates/products, **155**:34
Iron(II) porphyrins, reduction dehalogenation, **155**:22
Iron(II), visible spectra deuteroporphyrin IX, **155**:24
Irradiated Foods, Codex Alimentarius General Standard, **154**:3
Irradiation, classified as food additive, **154**:3
Irradiation, dairy products low temperature benefits, **154**:21
Irradiation, enhanced insect control with oxygen deficiency, **154**:17
Irradiation, grain nutrient stability, **154**:17
Irradiation, grain rancidity reduction, **154**:18
Irradiation, insect resistance, **154**:16
Irradiation, ionizing food preservation, **154**:1 ff.
Irradiation, meat packaging atmospheres, **154**:31
Irradiation, nitrite red meat pathogen control, **154**:27
Irradiation, red meat quality, **154**:27
Irradiation, starch breakdown to sugars, **154**:15
Irradiation, temperature effects insect control, **154**:16
Irrigation water, cadmium contamination, **154**:55 ff.
Isobacteriochlorins (Ni), environmental alkyl halide reactions, **155**:38
Isopods, relative tolerance to benomyl, **154**:123
I_w, dissolved metal influx marine mussels, **151**:49
Ixtoc-1 oil well, oil fractions fates, **153**:97
Ixtoc-1 oil well, oil spill, **153**:91

Jamaica Bay, New York, sediment trace metal study, **155**:97
Japan, organochlorine pesticide residues food, **152**:26

K_d (partition coefficient), metal dissolved/ marine mussels, **151**:53
K_d, atrazine (partition coefficient, soil:solution), **151**:126
Kelp, haloorganic synthesis, **155**:2
K_f, Fruendlich coefficients ethalfluralin soil, **153**:72
Kinetic model, trace element bioavailability marine mussels, **151**:54
Kinetic model, uncertainties metals marine mussels, **151**:57
Kinetic modeling, metal bioavailability marine mussels, **151**:39
Kingella kingae, as opportunistic pathogen, **152**:64
K_{oc} (organic carbon-water partition coefficient), defined, **158**:67
K_{oc}, trifluralin soil adsorption chart, **153**:19
K_{oc}s, polychlorinated alkanes, **158**:67
K_{ow}, trifluralin, **153**:5
K_{ow}s, polychlorinated alkanes, **158**:67
K_u, dissolved metal uptake rate constants marine mussels, **151**:48

L (lewisite), described, **156**:9
L, acute toxicity, **156**:103
L, carcinogenicity, **156**:107
L, ecotoxicology, **156**:143
L, environmental fate air, **156**:142
L, environmental fate soil, **156**:143
L, environmental fate water, **156**:142
L, estimated reference dose, **156**:108, 146, 151
L, genotoxicity, **156**:108
L, physical/chemical properties, **156**:15
L, subchronic toxicity, **156**:103, 109
L, toxicity mechanisms/symptoms, **156**:25
Lagadon rhomboides, rubber tire leachate toxicity, **151**:80, 89
Lake Erie drainage basin watershed, chlorpyrifos levels, **160**:41, 50, 97, 102
Lake sediments, alkyl halide probe responses, **155**:57

Landfills, scrap tire accumulation problems, **151**:70
Lanthanum, trace contamination estuaries, **155**:73 ff.
Latex, rubber tire leachates aquatic environment, **151**:67 ff.
LC_{50}, chlorpyrifos aquatic ecosystems, **160**:6
LC_{50}s, rice pesticides fish & aquatic organisms, **159**:106
LCC, see Low current configuration, **159**:117
LD_{50} values, GA, **156**:68
LD_{50} values, GB, **156**:79
LD_{50} values, GD, **156**:94
LDPE, chemical structure, **159**:3
LDPE, see Low density polyethylene, **159**:3
Leaching, atrazine 138, **151**:142
Lead, atmospheric measurements, **159**:26
Lead, bioaccumulation freshwater insect larvae, **158**:131
Lead, blood analytical methods, **159**:30
Lead, blood levels children, **159**:28
Lead, blood levels children vs dogs, **159**:37
Lead, blood levels dogs, **159**:35
Lead, blood levels general population, **159**:32
Lead, blood levels lead workers, **159**:31
Lead, children exposure sources, **159**:33
Lead, contamination Rio Grande Basin biota, **158**:10
Lead contamination, Uruguay, **159**:25 ff.
Lead, environmental Uruguay, **159**:28
Lead exposure, humans various routes, **159**:25
Lead exposure, lead workers, **159**:26
Lead exposure, occupational, **159**:26
Lead exposure, Uruguay sources, **159**:26
Lead, Gulf of Mexico coastal pollution, **157**:60
Lead industries, worker exposure, **159**:29
Lead inorganic, environmental human exposures, **159**:25
Lead, maximum permissible wastewater Mexico, **157**:60
Lead, occupational air levels, **159**:32
Lead, occupational exposure, **159**:29

Lead, organism levels Gulf of Mexico, **157**:74
Lead, pollution Gulf of Mexico, **157**:62
Lead, sediments Gulf of Mexico, **157**:66, 71
Lead, shark levels Gulf of Mexico, **157**:78
Lead, trace contamination estuaries, **155**:73 ff.
Lead, uptake marine mussels, **151**:48
Lead, workplace air levels, **159**:32
Legionella longbeachae, **152**:69
Legionella micdadei, **152**:69
Legionella, opportunistic pathogens drinking water, **152**:58
Legionella pneumophila, opportunistic water pathogen, **152**:69
Legionella pneumophila, oral infective dose humans, **152**:70
Legionella pneumophila, pneumonia causal pathogen, **152**:71
Legionella spp., occurrence in drinking water, **152**:69
LEL (lowest-effect level), nerve agent ChE inhibition, **156**:24
Lewisite (L), described, **156**:9
Lewisite, acute toxicity, **156**:103
Lewisite, carcinogenicity, **156**:107
Lewisite, developmental/reproductive effects, **156**:105, 109
Lewisite, ecotoxicology, **156**:143
Lewisite, environmental fate air, **156**:142
Lewisite, environmental fate soil, **156**:143
Lewisite, environmental fate water, **156**:142
Lewisite, estimated reference dose, **156**:108, 146, 151
Lewisite, genotoxicity, **156**:108
Lewisite, microbial degradation soil, **156**:140
Lewisite, physical/chemical properties, **156**:15
Lewisite, subchronic toxicity, **156**:103, 109
Lewisite, toxicity mechanisms/symptoms, **156**:25
Lewisite, visicant/systemic poison, **156**:25
Lichens, atmospheric mercury absorption, **157**:33
Lichens, biomonitors atmospheric pollution, **157**:4
Lindane, no-observed-effect concentrations soil fauna, **154**:94
Lindane, risk assessment soil fauna, **154**:91
Lindane, soil fauna field recovery, **154**:111, 120
Lindane, soil organism effects, **154**:94
Lipid oxidation, irradiated meat, **154**:27
Lipid peroxidation, mustard agent mechanisms, **156**:17
Lissorhoptrus oryzophilus, rice water weevil, **159**:97
Listeria monocytogenes, D-Values food irradiation, **154**:7, 24
LOAEL (lowest-observed-adverse-effect level), nerve agent ChE inhibition, **156**:24
Log K_{ow} values, PCDEs, **157**:136
Lolium perenne (grass), atmospheric biomonitor, **157**:6
Low current configuration (LCC), child cancer 117
Lumbricus terrestris, lindane effects, **154**:94
Luminescence, microbial pollutant assays, **159**:56
Lutetium, trace contamination estuaries, **155**:73 ff.
Lutra canadensis (Amer. River otter), PCB levels, **157**:105
Lutra lutra (Eurasian otter), PCBs effects, **157**:95 ff.
Lysimeter studies, atrazine, **151**:141

Magnetic field measurements, child brain tumors, **159**:114, 121, 123
Malathion, fish kill in California, **159**:100
Mammals, DDE residues Rio Grande Basin, **158**:17, 19
Mammals, polychlorinated alkanes toxicity, **158**:104, 106
Mammals, Rio Grande Basin, **158**:5
Mammals, scientific names (list), **158**:5
Manganese, bioaccumulation freshwater insect larvae, **158**:132
Manganese oxides, trace metal adsorption estuaries, **155**:88

Manganese, trace contamination estuaries, **155**:73 ff.
Mangroves, PAH crude oil spill effects, **153**:107
Maquiladora (assembly) plants, Rio Grande Basin, **158**:7
Marine bivalves, modeling metal bioavailability, **151**:39 ff.
Marine environment, PCDEs toxicity, **157**:131 ff.
Marine fish, PAH residues, **153**:106
Marine fish, rubber tire leachate toxicity, **151**:97
Marine mussels, as metal concentration monitors, **151**:43
Marine mussels, filtration rate, **151**:43
Marine mussels, metal assimilation from particulates, **151**:44
Marine mussels, metal bioavailability modeling, **151**:39 ff.
Marine mussels, metal influx dissolved phase, **151**:48
Marine mussels, trace element assimilation efficiency, **151**:44
Marine oil pollution, **153**:91 ff.
Marine oil spills, direct effects, **153**:110
Marine oil spills, indirect effects, **153**:110
Marine organisms, oil spill residues, **153**:105
Marine organisms, rubber tire leachate toxicity, **151**:86
Marine sediments, oil spill deposits, **153**:102
Market basket surveys, pesticide residues, **152**:2
Marone saxitilis, striped bass, **159**:99
Maximum residue limits (MRLs), organochlorines foods India, **152**:18
Mayflies (Ecdyonuridae), Pb, Zn, Cu vs sediment/water levels, **158**:140
Mayflies (Ephemeridae), larval bioaccumulation heavy metals, **158**:132
Mayflies (Ephemeridae), Pb, Zn, Cu vs sediment/water levels, **158**:138
Mayflies, trace metals vs sediment levels, **158**:134
Meat quality, packaging atmospheres, **154**:31

Mechanisms of toxicity, mustard agents, **156**:15
Mechanisms of toxicity, polychlorinated alkanes, **158**:112
Meningitis, opportunistic causal pathogens, **152**:71
Mercury, absorption by mosses/lichens via air, **157**:33
Mercury, air concentrations Brazil, **157**:30
Mercury, air pollution via vegetation burning, **157**:29
Mercury amalgamation process, typical losses, **157**:28
Mercury, Amazon contamination, 157:25 ff.
Mercury, aquatic animal biomonitors, **157**:35
Mercury, aquatic environment pollution, **157**:34
Mercury, bioaccumulation freshwater insect larvae, **158**:133
Mercury biomonitors, aquatic plants, **157**:34
Mercury, contamination Rio Grande Basin biota, **158**:10, 15
Mercury distillation, primary contamination source, **157**:26
Mercury, fish contamination levels Brazil, **157**:38
Mercury, Gulf of Mexico coastal pollution, **157**:56
Mercury, hair concentrations Amazon, **157**:42
Mercury, hair concentrations vs fish consumption, **157**:43
Mercury, human exposure Amazon, **157**:45
Mercury, human health, **157**:41
Mercury, organism levels Gulf of Mexico, **157**:73
Mercury, pollution air, **157**:29
Mercury, pollution Gulf of Mexico, **157**:62
Mercury, pollution terrestrial environment, **157**:29
Mercury, production in Mexico, **157**:59
Mercury, residues birds Rio Grande Basin, **158**:15, 18

Mercury, residues fish Rio Grande Basin, **158**:27

Mercury, residues invertebrates Rio Grande Basin, **158**:34

Mercury, residues plants Rio Grande Basin, **158**:35

Mercury, residues reptiles Rio Grande Basin, **158**:20

Mercury, residues Rio Grande water, **158**:41

Mercury, residues sediments Rio Grande Basin, **158**:36, 38

Mercury, Rio Grande Basin pollution, **158**:1 ff.

Mercury, river sediment contamination Amazon, **157**:37, 39

Mercury, sediments Gulf of Mexico, **157**:65

Mercury, shark levels Gulf of Mexico, **157**:78

Mercury, simple aquatic food chain, **157**:58

Mercury, soil contamination via air, **157**:31

Mercury, sources Amazon region, **157**:26

Mercury, sources/production Mexico, **157**:59

Mercury, storage biological compartments (illus), **157**:40

Mercury, terrestrial ecosystems levels, **157**:32

Mercury, trace contamination estuaries, **155**:73 ff.

Mercury, world production decreasing, **157**:58

Mesocosm studies, chlorpyrifos, **160**:78, 81

Metabolites, chlorpyrifos toxicity, **160**:57

Metal assimilation efficiency, factors affecting marine mussels, **151**:46

Metal bioavailability, kinetic model marine mussels, **151**:54

Metal bioavailability, marine mussels modeling, **151**:39 ff.

Metal concentrations, marine mussels model calculations, **151**:51

Metal efflux, marine mussels, **151**:50

Methane monooxygenase, dehalogenation processes, **155**:16

Methanobacter thermautrophicum, alkyl halide reduction, **155**:43, 46

Methanobacter thermautrophicum, dehalogenation products/yields, **155**:44

Methanogens, environmental alkyl halide reduction, **155**:42

Methodology, heavy metal bioaccumulation insect larvae, **158**:130

Methodology, oral reference dose derivation, **156**:26

Methomyl, no-observed-effect concentrations soil fauna, **154**:104

Methomyl, risk assessment soil fauna, **154**:91, 103

Methyl parathion, fish kill California, **159**:97

Methylene bromide, bacterial oxidation, **155**:11

Methylmercury, human hair exposure indicator, **157**:41

Methylmercury, human poisoning via fish, **157**:37

Methylnapthalene, rubber tire leachate component, **151**:79

Methylosinus trichosporium OB-3b, dehalogenation rate constants, **155**:15

Methylosinus trichosporium OB-3b, oxidation dehalogenation, **155**:10, 13

Methylumbelliferyl (MUF) substrates, microbial sediment asssay, **159**:59

MetPAD assay, toxic pollutant identification, **159**:63

MetPLATE assay, toxic pollutant identification, **159**:63

Mexico, coastal lead pollution, **157**:60

Mexico, coastal mercury pollution, **157**:56

Mexico, coastal metal pollution, **157**:53 ff.

Mexico, maximum permissible lead wastewater, **157**:60

Mexico, mercury sources, **157**:59

Microbial assay, microbial biomass/growth indicators, **159**:52

Microbial assay, microcosms sediments, **159**:50

Microbial assay, respiration/oxygen uptake methods, **159**:55
Microbial assay, sediment extracts, **159**:48
Microbial assay, sediment toxicity testing, **159**:52
Microbial assays, ATP methods, **159**:57
Microbial assays, luminescence pollutant methods, **159**:56
Microbial assays, microbial community structure/function, **159**:68
Microbial assays, microcalometry methods, **159**:57
Microbial assays, pollutant degradation studies, **159**:64
Microbial assays, sediment contaminants, **159**:41 ff.
Microbial assays, specific enzyme activity, **159**:58
Microbial hydrogenolysis, environmental alkyl halide dehalogenation, **155**:50
Microbial hydrolyses, environmental alkyl halides, **155**:45
Microbial methods, sediment contaminants, **159**:41 ff.
Microbial methods, sediment solid-phase contact, **159**:47
Microbial oxygen insertion, environmental alkyl halides, **155**:49
Microbial reductive elimination, environmental alkyl halides, **155**:50
Micrococcus radiodurans, radiation/temperature sensitivity, **154**:10
Microcosm studies, chlorpyrifos, **160**:78, 81
Microcosms, sediment microbial assays, **159**:50
Microorganisms, D-Values food irradiation, **154**:7
Microorganisms, rubber tire leachate toxicity, **151**:98
Microtox bioassay, sediment pollutants, **159**:56
Microtox Chronic Toxicity Test, **159**:56
Microtox® test, rubber tire leachate toxicity, **151**:93, 99
Milk, DDT/HCH residues India, **152**:9
Mineralization, atrazine, **151**:122

Minnesota Bioindicator study, described, **157**:10
Mississippi River drainage loads, trifluralin, **153**:26
Mixed-function oxygenase enzymes, polychlorinated alkane induction, **158**:95
Mode of action, EDTA as bactericide, **152**:98
Mode of action, EDTA as drug, **152**:99
Mode of action, ethalfluralin, **153**:66
Mode of action, trifluralin, **153**:2
Modeling, metal bioavailability marine mussels, **151**:39 ff.
Molecular structures, EDTA/DTPA, **152**:86
Molinate, rice herbicide fish kill California, **159**:96
Molinate, striped bass toxicity, **159**:98
Mollusks, heavy metals content Gulf of Mexico, **157**:74
Mollusks, lead content Gulf of Mexico, **157**:68
Molybdenum, bioaccumulation freshwater insect larvae, **158**:132
Moraxella, as opportunistic pathogen, **152**:64
Moraxella nonliquefaciens, respiratory/eye infections, **152**:64
Moraxella, occurrence drinking water, **152**:66
Moraxella, opportunistic pathogens drinking water, **152**:58
Moraxella osloensis, as opportunistic pathogen, **152**:64
Moraxella spp., radiation sensitivity food, **154**:8
Morpholinothio-benzothiazole, new rubber tire leachate, **151**:78, 106
Morpholinyl-benzothiazole, street runoff toxicity, **151**:106
Mosquito control, organochlorine insecticides Africa, **151**:2
Moss, heavy metal tolerance, **157**:11
Moss, metal concentrations Europe transect, **157**:10
Mosses, atmospheric mercury absorption, **157**:33
Mosses, biomonitors atmospheric pollution, **157**:3

Motile bacteria, pollutant assay, **159:**62
MRLs (maximum residue limits), organochlorines foods India, **152:**18
Mucopolysaccharide matrices, microbial bound water, **155:**124
MUF, see Methylumbelliferyl substrates, **159:**59
Mussel (*Mytilus edulis*), cadmium indicator, **154:**65
Mussel Watch Programs, **151:**43
Mussels, marine metal concentrations, **151:**39 ff.
Mussels, marine oil spill residues, **153:**107
Mussels, metal assimilation from particulates, **151:**44
Mustard agents, ecotoxicology, **156:**132
Mustard agents, environmental fate air, **156:**128
Mustard agents, environmental fate/effects, **156:**127
Mustard agents, environmental fate soil, **156:**131
Mustard agents, environmental fate water, **156:**128
Mustard agents, estimated reference dose, **156:**146
Mustard agents, fish effects, **156:**132
Mustard agents, mechanisms of toxicity, **156:**16
Mustard agents, physical/chemical properties, **156:**12
Mustard agents, toxicity symptoms, **156:**15
Mustard gases, described, **156:**9
Mustard gases, estimated reference dose, **156:**146
Mustard gases, physical/chemical properties, **156:**10, 12
Mustela putorius furo (ferret), PCB sensitivity, **157:**108
Mutagenic effects, cadmium, **154:**68
Mutagenicity, polychlorinated alkanes, **158:**114
Mutagenicity tests, rubber tire leachate, **151:**92
Mycobacteria, occurrence in drinking water, **152:**68

Mycobacterium avium complex, composed of 28 seovars, **152:**66
Mycobacterium avium complex, occurrence in drinking water, **152:**66, 68
Mycobacterium avium, oral infective dose humans, **152:**67
Mycobacterium avium, septicemia causal pathogen, **152:**72
Mycobacterium intracellulare, as opportunistic pathogen, **152:**66
Mycobacterium, opportunistic pathogens drinking water, **152:**58
Mycobacterium scrofulaceum, as opportunistic pathogen, **152:**66
Myoglobin, alpha-carbon skeleton (illus.), **155:**27
Mytilus edulis (mussel), cadmium indicator, **154:**65
Mytilus edulis, metal assimilation efficiency, **151:**44
Mytilus edulis, metal assimilation from particulates, **151:**44
Mytilus edulis, metal bioavailability modeling, **151:**39 ff.
Mytilus galloprovincialis, metal bioavailability, **151:**51

N-dealkylation, atrazine, **151:**122
NAFTA (North American Free Trade Agreement), **158:**6
Naphthalene, rubber tire leachate component, **151:**79
Narragansett Bay, Rhode Island, sediment trace metal study, **155:**97
National Water Quality Assessment Program (NAWQA), chlorpyrifos, **160:**31
National Water Quality Assessment Program, water sampling locations US, **160:**34
Natural organic matter, bound-water, **155:**115 ff.
Natural organic matter, characterization, **155:**116
Natural organic matter, conceptual illustration, **155:**118
Natural organic matter, dissolved organics, **155:**116
Natural organic matter, membrane filtration illustration, **155:**119

Natural organic matter, nonfilterable particulates, **155**:116
Natural organic matter-bound water, defined, **155**:121
Natural organic matter-bound water, environmental significance, **155**:125
Natural rubber, automobile tire composition, **151**:68
NAWQA (National Water Quality Assessment Program), chlorpyrifos, **160**:31
NAWQA, water sampling locations US, **160**:34
Neodymium, trace contamination estuaries, **155**:73 ff.
Neomysis mercedis, opposum shrimp, **159**:98
Nerve agent ChE inhibition, LEL, **156**:24
Nerve agent ChE inhibition, LOEL, **156**:24
Nerve agent ChE inhibition, NOEL, **156**:24
Nerve agents, acethycholinesterase inhibitions, **156**:18
Nerve agents, blood cholinesterase effects, **156**:20
Nerve agents, delayed neuropathy, **156**:19
Nerve agents, ecotoxicology, **156**:141
Nerve agents, environmental fate, **156**:133
Nerve agents, environmental fate air, **156**:134
Nerve agents, environmental fate soil, **156**:138
Nerve agents, estimated reference dose, **156**:146
Nerve agents, fish toxicity, **156**:141
Nerve agents, hydrolysis to alkyl methylphosphonates, **156**:133
Nerve agents, microbial degradation soil, **156**:140
Nerve agents, nervous system effects, **156**:19
Nerve agents, physical/chemical properties, **156**:13
Nerve agents, toxicity mechanisms/symptoms, **156**:17
Nerve gases, chemical warfare, described, **156**:9
Nervous system effects, nerve agents, **156**:19

Neurotoxicity, GA, **156**:72
Neurotoxicity, GB, **156**:87
Neurotoxicity, GD, **156**:98
Neurotoxicity, VX, **156**:59
Neutron activation analysis, plant atmospheric biomonitors, **157**:8
New Zealand, organochlorine pesticide residues food, **152**:30
Nickel, bioaccumulation freshwater insect larvae, **158**:132
Nickel, contamination Rio Grande Basin biota, **158**:10
Nickel isobacteriochlorins, alkyl halide reaction rate constants, **155**:42
Nickel isobacteriochlorins, alkyl halide reactions, **155**:38
Nickel isobacteriochlorins, *n*-butyl chloride reaction spectrum, **155**:41
Nickel, organism levels Gulf of Mexico, **157**:77
Nickel, pollution Gulf of Mexico, **157**:65
Nickel, sediments Gulf of Mexico, **157**:69
Nickel, trace contamination estuaries, **155**:73 ff.
Nitrite, irradiation red meat pathogen control, **154**:27
Nitrogen mustard (HN2), described, **156**:9
Nitrogen mustard (HN2), physical/chemical properties, **156**:11
Nitrogen mustards, hydrolysis, **156**:130
NMR (^{13}C) spectra, bacterial dehalogenation, **155**:14
No-observed-effect concentrations, pesticides soil fauna, **154**:83 ff.
No-till herbicide leaching, atrazine, **151**:143
NO_3, atrazine groundwater correlation, **151**:136
NOEL (no-observed-effect-level), nerve agent ChE inhibition, **156**:24
Non-stockpile chemical materiel, defined, **156**:2
Nonfermentative gram-negative bacteria, as opportunistic pathogens, **152**:70
Nontarget organisms, pesticides soil fauna, **154**:83 ff.
North American Free Trade Agreement (NAFTA), **158**:6

Nosocomial (hospital-acquired) pathogens, **152**:59

Occupational exposure, lead, **159**:29
Ocean oil pollution, **153**:91 ff.
Off-flavors, irradiated dairy products, **154**:21
OHCC, see Ordinary high current configuration, **159**:114
Oil pollution, Caribbean Sea, **153**:91 ff.
Oil pollution, Gulf of mexico, **153**:91 ff.
Oil pollution, marine, **153**:91 ff.
Oil spills, coral reef effects, **153**:109
Oil spills, effects marine organisms, **153**:107
Oil spills, Gulf of Mexico, **153**:91 ff.
Oil spills, high risk areas Gulf of Mexico, **153**:96
Oil spills, major Caribbean Region, **153**:97
Oil spills, mangrove effects, **153**:108
Oil spills, marine direct/indirect effects, **153**:110
Oil spills, marine organism effects, **153**:105
Oil spills, marine sediments, **153**:102
Oil spills, seeps marine, **153**:98
Oil spills, sources, **153**:92, 94
Oil tanker trade lanes, Gulf of Mexico, **153**:93
Oil tanker transit frequency, Carribean Sea, **153**:95
OLCC, see Ordinary low current configuration, **159**:117
Onchorynchus mykiss, rubber tire leachate toxicity, **151**:83, 87, 94
Onchorynchus tshawytscha, chinook salmon, **159**:99
Opportunistic bacterial pathogens, drinking water, **152**:57 ff.
Opportunistic bacterial pathogens drinking water, importance ranking, **152**:58
Opportunistic drinking water pathogens, ranking of importance, **152**:71
Opposum shrimp, *Neomysis mercedis*, **159**:99
Oral reference doses, chemical warfare agents, **156**:1 ff.

Oral reference doses, derivation methodology, **156**:26
Ordinary high current configuration (OHCC), child cancer, **159**:114
Ordinary low current configuration (OLCC), child cancer, **159**:117
Organelles, rubber tire leachate toxicity, **151**:98
Organic arsenical warfare chemicals, described, **156**:9
Organic carbon, atrazine soil sorption, **151**:125
Organic carbon-water partition coefficient (K_{oc}), defined, **158**:67
Organic compounds, residues Rio Grande water, **158**:4
Organic halides, synthesis in sea, **155**:2
Organic matter, trace metal adsorption estuaries, **155**:85
Organochlorine insecticide residues, African fauna, **151**:1 ff.
Organochlorine insecticide use, Africa, **151**:2
Organochlorine insecticides, adverse effects fauna/flora Africa, **151**:3
Organochlorine insecticides, African publications, **151**:4
Organochlorine insecticides, chemical names (list), **158**:7
Organochlorine insecticides, contamination Rio Grande Basin biota, **158**:10
Organochlorine insecticides, history, **151**:1
Organochlorine insecticides, interference polychlorinated alkanes analyses, **158**:74
Organochlorine insecticides, residues Rio Grande water, **158**:40
Organochlorine pesticide ingestion, country comparisons, **152**:36
Organochlorine pesticide residues, food Asia, **152**:1 ff.
Organochlorine pesticide residues, food Asia-Pacific region, **152**:1 ff.
Organochlorine pesticide residues food, Australia, **152**:30
Organochlorine pesticide residues food, country comparisons, **152**:27, 34
Organochlorine pesticide residues food, Indonesia, **152**:24

Organochlorine pesticide residues food, New Zealand, **152**:30
Organochlorine pesticide residues, food Oceanic countries, **152**:1 ff.
Organochlorine pesticide residues food, Oceanic countries, **152**:28
Organochlorine pesticide residues food, Southeast Asia, **152**:19
Organochlorine pesticides, dietary intake Asia-Pacific region, **152**:35
Organochlorine pesticides, dietary intake Australia, **152**:42
Organochlorine pesticides, dietary intake India, **152**:40
Organochlorine pesticides, dietary intake Thailand, **152**:41
Organochlorine pesticides, dietary intake, various countries, **152**:43
Organochlorine residues, aquatic invertebrates Africa, **151**:13, 17
Organochlorine residues, aquatic vertebrates Africa, **151**:16
Organochlorine residues, bird eggs Africa, **151**:15,24
Organochlorine residues, birds Africa, **151**:8, 12
Organochlorine residues, fish Africa, **151**:18, 26
Organochlorine residues, invertebrates Africa, **151**:9, 13
Organochlorine residues, mammals Africa, **151**:10, 13
Organochlorine residues, reptiles/amphibians Africa, **151**:8
Organochlorine residues, temporal faunal changes Africa, **151**:16
Organochlorine residues, terrestrial invertebrates Africa, **151**:6, 9
Organophosphate insecticide, chlorpyrifos risk assessment, **160**:1 ff.
Organophosphate insecticides, toxicity soil arthropods, **154**:123
Organophosphate-induced delayed neuropathy, GB, **156**:88
Otters, PCBs in feces, **157**:104
Otters, PCBs suspected in population declines, **157**:95 ff.
Oxidation, atmospheric halocarbons, **155**:9

Oxidation, environmental dehalogenation processes, **155**:9
Oxidation, polychlorinated alkanes, **158**:68
Oysters, heavy metal levels Gulf of Mexico, **157**:75
Oysters, heavy metal levels Mexican Pacific, **157**:82
Oysters, oil spill residues, **153**:106
Oysters, PAH residues, **153**:106
Ozonation, EDTA sensitivity, **152**:88

P-450 enzymes, dehalogenation oxidation rates, 155:36
Pacific Islands, organochlorine pesticide residues food, **152**:29
PAHs, described in marine sediments, **153**:104
PAHs, marine sediment oil spills, **153**:103
PAHs, residues oysters fish, **153**:106
PAHs, rubber tire fire smoke, **151**:70
PAHs, rubber tire leachate component, **151**:79
Palaemonetes pugio, rubber tire leachate toxicity, **151**:90
Parathion-ethyl, no-observed-effect concentrations soil fauna, **154**:96
Parathion-ethyl, risk assessment soil fauna, **154**:91, 97, 99
Parathion-ethyl, soil fauna field recovery, **154**:113, 120
Particulate organic matter (POM), (illus.), **155**:118
PATRIOT, pesticide transport model, **153**:86
PCB concentrations, vary with analytical method, **157**:104
PCB concentrations, vary with otter populations, **157**:102
PCB congener patterns, European otters, **157**:101
PCB congeners, metabolism, **157**:97
PCB congeners, otter differential metabolism, **157**:116
PCB isomers, metabolism, **157**:97
PCB residues, food Asia, **152**:1 ff.
PCB residues, food Oceanic countries, **152**:1 ff.

PCBs, concentration declines with time otters, **157**:103
PCBs, concentration vs age otters, **157**:103
PCBs, concentration vs otter body condition, **157**:109
PCBs, concentration vs otter death cause, **157**:110
PCBs, concentration vs otter population performance, **157**:110
PCBs, concentration vs otter reproductive status, **157**:110
PCBs, concentrations European otters, **157**:96, 98
PCBs, contamination Rio Grande Basin biota, **158**:10
PCBs, effects American mink (*Mustela vison*), **157**:107
PCBs, effects Eurasian otter (*Lutra lutra*), **157**:95 ff.
PCBs, effects European otters, **157**:107
PCBs, methyl sulfonyl metabolites otters, **157**:102
PCBs, otter feces, **157**:104
PCBs, pathological effects European otter, **157**:108
PCBs, pathological effects seals, **157**:108
PCBs, proposed safe levels sediment/fish/otter, **157**:114
PCBs, residues food Australia/New Zealand, **152**:30
PCBs, residues food India, **152**:15,16
PCBs, residues food Thailand, **152**:23
PCBs, rubber tire leachate, **151**:73
PCBs, see also polychlorinated biphenyls, **157**:95 ff.
PCBs, suspected Eurasian otter decline, **157**:96
PCDDs, see also polychlorinated dibenzo-*p*-dioxins, **157**:131
PCDEs, acute toxicity trout, **157**:139
PCDEs, analytical methods, **157**:134
PCDEs, chemical structure (illus.), **157**:131
PCDEs, concentrations fresh water fish, **157**:137
PCDEs, concentrations marine organisms, **157**:138

PCDEs, congeners in technical chlorophenol, **157**:133
PCDEs, distribution marine environment, **157**:136
PCDEs, environmental sources, **157**:131
PCDEs, enzyme induction, **157**:139
PCDEs, immunosupression, **157**:139
PCDEs, Log K_{ow} values, **157**:136
PCDEs, no-observable-effect levels rats, **157**:140
PCDEs, photodechlorination, **157**:137
PCDEs, presence in chlorophenols, **157**:131
PCDEs, see also polychlorinated diphenly-ethers, **157**:131
PCDEs, TCDD (dioxin) activity, **157**:141
PCDEs, tetrachlorodibenzo-*p*-dioxin (TCDD) activity, **157**:141
PCDEs, toxicity in marine environment, **157**:131 ff.
PCDFs, see also polychlorinated dibenzofurans, **157**:131
PCR, see Polymerase chain reaction technology, **159**:69
Perlodidae (Stoneflies), trace metals vs sediment levels, **158**:134
Pesticide degradation, rice field water holding times, **159**:97
Pesticide effects, aquatic ecosystems, **159**:95 ff.
Pesticide effects, soil fauna recovery, **154**:83 ff.
Pesticide risk assessment, soil fauna test designs,85
Pesticide runoff simulation model, ANSWERS, **153**:23
Pesticide transport modeling, ethalfluralin, **153**:85
Pesticides, African import values, **151**:4
Pesticides, estimated ecotoxicologic recovery time soil fauna, **154**:109
Pesticides, no-observed-effect concentrations soil fauna, **154**:83 ff.
Pesticides, predicted environmental concentrations, **154**:83
Pesticides, predicted no-effect concentrations soil fauna, **154**:83
Pesticides, ranking of effects soil fauna, **154**:110

Pesticides, risk assessment ecotoxicologic recovery time, **154:**108
Pesticides, soil fauna predicted vs observed recovery, **154:**118
Pesticides, sublethal effects soil fauna, **154:**83 ff.
PETE, chemical structure, **159:**3
PETE, see Polyethylene terephthalate, **159:**3
Petrochemical plastics, chemical structures, **159:**3
Petroleum hydrocarbons, marine organisms, **153:**105
Petroleum hydrocarbons, marine samples Gulf of Mexico, **153:**100
Petroleum pollution, Caribbean Sea, **153:**91 ff.
Petroleum pollution, Gulf of Mexico, **153:**91 ff.
pH, atrazine soil adsorption effects, **151:**128
PHA, see Polyhydroxyalkanoate, **159:**2
PHAs, biodegradation, **159:**17
PHAs, biopolymer class, **159:**2
PHAs, biosynthesis, **159:**7
PHAs, biosynthesized polyesters, **159:**4
PHAs, carbon cycle, **159:**19
PHAs, microorganism synthesis, **159:**7
PHAs, production transgenic insect cells, **159:**15
PHAs, production transgenic plants, **159:**14
PHAs, transgenic production, **159:**13
PHAs, transgenic production *Escherichia coli*, **159:**14
PHB, advantages over traditional plastics, **159:**21
PHB, biodegradation, **159:**17
PHB, biosynthetic pathway, **159:**10
PHB, degradation pathway *Alcaligenes eutrophus*, **159:**18
PHB, enzymatic degradation (illus.), **159:**9
PHB, industrial production, **159:**12
PHB, microorganism synthesis/degradation, **159:**3
PHB polymerization, **159:**10
PHB, see Poly-R-3-hydroxybutyrate, **159:**3

PHB, stereo chemical structures, **159:**8
PHB synthase, **159:**12
PHB synthesis, natural/engineered pathways, **159:**16
PHB, synthesized by microorganisms, **159:**3
PHBV, biosynthetic pathway, **159:**11
PHBV, chemical structure, **159:**6
PHBV, physical properties, **159:**6
PHBV, see Polyhydroxybutyrate-co-valerate, **159:**6
Phosphofluoridate fluorohydrolase (sarinase), **156:**21
Phosphorylation, AChE chlorpyrifos (diag), **160:**13
Photobacterium phosphoreum, (now *Vibrio fisheri*), **159:**56
Photobacterium phosphoreum, inhibition rubber tire leachate, **151:**93, 99
Photochemical degradation, EDTA, **152:**93
Photodechlorination, PCDEs, **157:**137
Photodecomposition, ethalfluralin, **153:**67
Photodecomposition, trifluralin, **153:**6
Photodegradable plastics, **159:**2
Photodegradation, nerve agents, **156:**134
Photohydrolysis, environmental dehalogenation processes, **155:**6
Photolysis, ethalfluralin, **153:**67
Photolysis, GD, VX, **156:**134
Photolysis, humic acid inhibits EDTA, **152:**88
Photolysis, nerve agents, **156:**134
Photolysis, polychlorinated alkanes, **158:**68
Photolysis, trifluralin, **153:**6
Physical/chemical properties, chemical warfare agents, **156:**10, 12, 14, 16
Physical/chemical properties, chlorpyrifos, **160:**12
Physical/chemical properties, polychlorinated alkanes, **158:**63
Physical properties, polyhydroxyalkanoates, **159:**4
Physicochemical properties, chlorpyrifos, **160:**12
Physicochemical properties, ethalfluralin, **153:**67, 88

Physicochemical properties, polychlorinated alkanes, **158**:115
Physicochemical properties, trifluralin, **153**:5, 88
Picea (spruce) tree species, atmospheric biomonitor, **157**:5
Pimephales promales, fathead minnow, **159**:99
Pimephales promales, rubber tire leachate toxicity, **151**:83, 87
Pinus (pine) tree species, atmospheric biomonitor, **157**:5
Plant atmospheric biomonitors, taxa choice, **157**:13
Plant biomonitors, air pollution, **157**:1 ff.
Plants, air pollution biomonitors, **157**:1 ff.
Plants, scientific names (list), **158**:6
Plasma/atomic emission spectrophotometry, plant atmospheric biomonitors, **157**:8
Plasma-ChE (butyrylcholinesterase), nerve agent effects, **156**:20
Plastics (petrochemical), chemical structures, **159**:3
Plastics, biodegradable via starch, **159**:2
Plastics, biosynthesized polymers, **159**:2
Plastics, degradation properties, **159**:20
Plastics, household volume in landfills, **159**:1
Plastics, incineration energy utilization, **159**:2
Plastics, photodegradation by UV, **159**:2
Plastics, recycling numbering system, **159**:2
Plastics, waste disposal patterns, **159**:1
Platichthyus stellatus, rubber tire leachate toxicity, **151**:80
Plesiomonas shigelloides, D-Values food irradiation, **154**:7
Pleurozium schreberi (moss), atmospheric biomonitor, **157**:4, 13
Pneumonia, opportunistic causal pathogens, **152**:71
Poa annua (grass), atmospheric biomonitor, **157**:6
Poecilia reticulata, rubber tire leachate toxicity, **151**:77, 88
Point-source monitoring, plant atmospheric biomonitors, **157**:9

Pollution biomonitors, plants, **157**:1 ff.
Pollution, plants as biomonitors, **157**:1 ff
Pollution regulation, plant biomonitors, **157**:1 ff.
Poly(ADP-ribose)polymerase, mustard agent mechanism, **156**:17
Poly-R-3-hydroxybutyrate (PHB), synthesized by microorganisms, **159**:3
Polybutadiene rubber, chemistry, **151**:69
Polychlorinated alkanes, analytical methods, **158**:71
Polychlorinated alkanes, analytical quantitation, **158**:72
Polychlorinated alkanes, aquatic organisms contamination, **158**:77, 83
Polychlorinated alkanes, aquatic toxicity, **158**:97
Polychlorinated alkanes, atmospheric concentrations, **158**:88
Polychlorinated alkanes, bioaccumulation, **158**:88
Polychlorinated alkanes, bioconcentration factors, **158**:91
Polychlorinated alkanes, biodegradation, **158**:68
Polychlorinated alkanes, biomagnification factors, **158**:93
Polychlorinated alkanes, biotransformation, **158**:93
Polychlorinated alkanes, bird toxicity, **158**:104, 106
Polychlorinated alkanes, carcinogenicity, **158**:114
Polychlorinated alkanes, embryo toxicity, **158**:105
Polychlorinated alkanes, environmental fate modeling, **158**:115
Polychlorinated alkanes, environmental levels, **158**:76
Polychlorinated alkanes, enzyme induction, **158**:93
Polychlorinated alkanes, exposure/risk assessment, **158**:115
Polychlorinated alkanes, extraction/isolation, **158**:71
Polychlorinated alkanes, fish toxicity, **158**:100
Polychlorinated alkanes, flame retardant use, **158**:54

Polychlorinated alkanes, formula group abundance profiles, **158**:75
Polychlorinated alkanes, fugacity II model calculations, **158**:115
Polychlorinated alkanes, GCMS analysis, **158**:72
Polychlorinated alkanes, half-lives aquatic biota, **158**:94
Polychlorinated alkanes, Henry's Law Constants, **158**:65
Polychlorinated alkanes, HPLC total ion chromatogram, **158**:58
Polychlorinated alkanes, hydrolysis, **158**:68
Polychlorinated alkanes, industrial formulations, **158**:57
Polychlorinated alkanes, industrial synthesis, **158**:55
Polychlorinated alkanes, invertebrate toxicity, **158**:98
Polychlorinated alkanes, K_{oc}s, **158**:67
Polychlorinated alkanes, K_{ow}s, **158**:66
Polychlorinated alkanes, mammalian toxicity, **158**:104, 106
Polychlorinated alkanes, mechanisms of toxicity, **158**:112
Polychlorinated alkanes, microbial toxicity, **158**:97
Polychlorinated alkanes, mutagenicity, **158**:114
Polychlorinated alkanes, number of positional isomers (table), **158**:59
Polychlorinated alkanes, oxidation, **158**:68
Polychlorinated alkanes, photolysis, **158**:68
Polychlorinated alkanes, physicochemical properties, **158**:115
Polychlorinated alkanes, plant toxicity, **158**:98
Polychlorinated alkanes, release into environment, **158**:61
Polychlorinated alkanes, reproductive toxicity, **158**:105
Polychlorinated alkanes, residues birds, **158**:85
Polychlorinated alkanes, residues foodstuffs, **158**:85, 87
Polychlorinated alkanes, residues humans, **158**:85, 87
Polychlorinated alkanes, residues mammals, **158**:85
Polychlorinated alkanes, risk assessment, **158**:117
Polychlorinated alkanes, sediment bioavailability, **158**:96
Polychlorinated alkanes, sediment contamination, **158**:77, 81
Polychlorinated alkanes, sewage sludge contamination, **158**:77, 81
Polychlorinated alkanes, sublethal toxicity indicators, **158**:112
Polychlorinated alkanes, teratogenicity, **158**:105
Polychlorinated alkanes, thermal degradation, **158**:70
Polychlorinated alkanes, toxicity, **158**:96, 98
Polychlorinated alkanes, toxicokinetics, **158**:88
Polychlorinated alkanes, uses, **158**:54, 60
Polychlorinated alkanes, vapor pressures, **158**:64
Polychlorinated alkanes, water contamination, **158**:76, 78
Polychlorinated alkanes, water solubility, **158**:63
Polychlorinated alkanes, world consumption, **158**:60
Polychlorinated biphenyl (PCB) residues, food Asia, **152**:1 ff.
Polychlorinated biphenyls, see also PCBs, **157**:95 ff.
Polychlorinated dibenzo-p-dioxins, see also PCDDs, **157**:131
Polychlorinated dibenzofurans, see also PCDFs, **157**:131
Polychlorinated diphenyl ethers (PCDEs), environmental sources, **157**:131
Polychlorinated diphenylethers (PCDEs), chemical structure, **157**:131
Polychlorinated diphenylethers, see also PCDEs, **157**:131
Polychlorinated n-alkanes, environmental chemistry/toxicology, **158**:53 ff.
Polycyclic aromatic hydrocarbons, see PAHs, **153**:103
Polyethylene high-density (HDPE), chemical structure, **159**:3

Polyethylene low-density (LDPE), chemical structure, **159**:3
Polyethylene plastics, recycling frequency, **159**:2
Polyethylene terephthalate (PETE), chemical structure, **159**:3
Polyhydroxyalkanoates (PHAs), biopolymer class, **159**:2
Polyhydroxyalkanoates, biosynthesis, **159**:7
Polyhydroxyalkanoate, general chemical structure, **159**:4
Polyhydroxyalkanoates, physical properties, **159**:4
Polyhydroxyalkanoates, stereochemistry, **159**:7
Polyhydroxybutyrate-co-valerate (PHBV), chemical structure, **159**:6
Polyhydroxybutyrate, biosynthetic pathway, **159**:10
Polyhydroxybutyrate, chemical structure, **159**:4
Polyhydroxybutyrate, degraded by microorganisms, **159**:1 ff.
Polyhydroxybutyrate, physical properties, **159**:5
Polyhydroxybutyrate, stereo chemical structures, **159**:8
Polyhydroxybutyrate, synthesized by microorganisms, **159**:1 ff.
Polyhydroxyvalerate, chemical structure, **159**:4
Polymerase chain reaction (PCR) technology, **159**:69
Polymerization, PHB by microorganisms, **159**:10
Polypropylene (PP), chemical structure, **159**:3
Polypropylene, #5 grade recyclable plastic, **159**:4
Polypropylene, physical properties, **159**:5
Polystyrene (PS), chemical structure, **159**:3
Polytrichum formosum (moss), atmospheric biomonitor, **157**:4, 13
Polyvinyl chloride (PVC), chemical structure, **159**:3
Populas nigra (black cottonwood tree), atmospheric biomonitor, **157**:6

Population modeling, environmental risk assessment chlorpyrifos, **160**:110
Pore waters, sediment microbial assay, **159**:48
Pork, *Trichinella* irradiation control, **154**:25
Poultry meat, irradiation preservation, **154**:32
PP (polypropylene), chemical structure, **159**:3
Primitive neuroectodermal tumor, child brain tumors, **159**:112
Probabilistic risk assessment, chlorpyrifos, **160**:84, 96
Procambarus clarkii (crayfish), cadmium indicator, **154**:65
PS (polystyrene), chemical structure, **159**:3
Pseudomonas aeruginosa, as opportunistic pathogen, **152**:59
Pseudomonas aeruginosa, infection risk drinking water, **152**:74
Pseudomonas aeruginosa, meningitis causal pathogen, **152**:72
Pseudomonas aeruginosa, occurrence bottled/tap water, **152**:60
Pseudomonas aeruginosa, oral infective dose humans, **152**:61
Pseudomonas aeruginosa, persons at infection risk, **152**:59
Pseudomonas aeruginosa, pneumonia causal pathogen, **152**:71
Pseudomonas aeruginosa, septicemia causal pathogen, **152**:72
Pseudomonas lemoignei, PHA biodegradation, **159**:17
Pseudomonas oleovorans, PHA accumulation, **159**:7
Pseudomonas, opportunistic pathogens drinking water, **152**:58
Pseudomonas putida (Ppg-786), cytochrome P-450, **155**:17
Pseudomonas putida, dehalogenation processes, **155**:30, 33
Pseudomonas, radiation/temperature sensitivity, **154**:10
Pseudomonas stutzeri, PHA biodegradation, **159**:17
Pseudoschleropodium purum (moss), atmospheric biomonitor, **157**:4, 13

Pulsed exposures, chlorpyrifos toxicity fish, **160**:58
PVC, chemical structure, **159**:3
PVC, see Polyvinyl chloride, **159**:3

R2A bacteria culture medium, **152**:58
RAD (radiation absorbed dose), defined, **154**:2
Radappetization (food irradiation), defined, **154**:4
Radiation absorbed dose (RAD), defined, **154**:2
Radiation chemistry, in food irradiation, **154**:4
Radiation, ionizing food preservation, **154**:1 ff.
Radicidation (food irradiation), defined, **154**:4
Radiobiology, in food irradiation, **154**:4
Radiosensitivity, chromosome volume correlation, **154**:5
Radurization (food irradiation), defined, **154**:4
Rain, atrazine contamination, **151**:118
Rainbow trout, rubber tire leachate toxicity, **151**:83, 87, 94
Rancidity, irradiated meat packaging, **154**:31
Rancidity, irradiation reduction grains, **154**:18
RBC-AChE (red blood cell acetylcholinesterase), nerve agent effects, **156**:20
RBC-ChE activity, species variation, **156**:22
Reciprocity, chlorpyrifos time vs toxicity fish, **160**:59
Recycling, rubber tires, **151**:72
Red meat, irradiated quality, **154**:27
Red meat pathogens, D_{10} irradiation values, **154**:24
Red meat quality, packaging atmospheres, **154**:31
Reducing enzymes, environmental dehalogenation processes, **155**:35
Reduction, environmental dehalogenation processes, **155**:7, 18
Reference dose (RfD), chemical warfare agents, **156**:8
Reference dose (RfD), defined, **156**:26

Refugia, risk assessment chlorpyrifos, **160**:92
Reproductive toxicity, polychlorinated alkanes, **158**:105
Reptiles, DDE residues Rio Grande Basin, **158**:19
Reptiles, mercury residues Rio Grande Basin, **158**:20
Reptiles, organochlorine residues Africa, **151**:8, 10
Reptiles, Rio Grande Basin, **158**:5
Reptiles, scientific names (list), **158**:5
Reptiles, selenium residues Rio Grande Basin, **158**:20
Residues, ethalfluralin decline soil, **153**:77
Residues, ethalfluralin soil, **153**:82
Residues, ethalfluralin vegetation, **153**:82
Residues, nonextractable soil-bound ethalfluralin, **153**:70
Residues, nonextractable soil-bound trifluralin, **153**:15
Residues, organochlorine pesticides foods Asia, **152**:1 ff.
Residues, PCBs foods Asia, **152**:1 ff.
Respiration/oxygen uptake methods, microbial assays, **159**:55
Return frequency, risk assessment chlorpyrifos, **160**:92, 109
RfD (reference dose), defined, **156**:26
RfDs, chemical warfare agents, **156**:8, 146
Rhodanese, cyanide conversion to thiocyanate, **156**:26
Rhytidiadelphus squarrosus (moss), atmospheric biomonitor, **157**:4, 13
Rice pesticide effects on microinvertebrate fish prey, **159**:108
Rice pesticide LC_{50}s to fish & aquatic organisms, **159**:106
Rice pesticides, aquatic toxicity, **159**:106
Rice pesticides, effects aquatic ecosystems, **159**:95 ff.
Rice pesticides, impacts California aquatic ecosystems, **159**:104
Rice water weevil, *Lissorhoptrus oryzophilus*, **159**:97
Rio Bravo Basin, DDE pollution, **158**:1 ff.

Rio Bravo Basin, metal pollution, **158**:1 ff.
Rio Grande Basin, DDE pollution, **158**:1 ff.
Rio Grande Basin, heavy metal pollution, **158**:1 ff.
Rio Grande Basin, map, **158**:2
Rio Grande Basin, mercury pollution, **158**:1 ff.
Rio Grande Basin, pollution sampling stations (1992–93), **158**:24
Rio Grande Basin, selenium pollution, **158**:1 ff.
Rio Grande River, most endangered river North America, **158**:7
Rio Grande River, tributaries, **158**:1
Ripening inhibition, fruit irradiation, **154**:14
Risk assessment, chlorpyrifos aquatic environments, **160**:1 ff.
Risk assessment, opportunistic bacterial pathogens water, **152**:57 ff.
Risk assessment, opportunistic drinking water pathogens, **152**:72
Risk assessment, pesticide effects soil fauna, **154**:83 ff.
Risk assessment, polychlorinated alkanes, **158**:117
Risk characterization, chlorpyrifos aquatic environments, **160**:11
Risk factors, child brain tumors, **159**:113
River basin drainage loads, trifluralin, **153**:26
River water, organic compound contaminants Rio Grande, **158**:40
River water, organochlorine contaminants Rio Grande, **158**:40
Rivers, trace metal levels, **155**:79
Roadbeds, scrap rubber tire utilization, **151**:76
Rubber, activators manufacturing, **151**:70
Rubber, antioxidants manufacturing, **151**:70, 74
Rubber chemistry, **151**:69
Rubber debris, atmospheric deposition/runoff, **151**:107
Rubber, natural chemistry, **151**:69
Rubber, synthetic chemistry, **151**:69
Rubber tire airborne fragments, health effects, **151**:81

Rubber tire burial, environmental problems, **151**:76
Rubber tire dust, composition, **151**:81
Rubber tire leachate, chemical characterization, **151**:73
Rubber tire leachates, biological effects, **151**:80 ff.
Rubber tire leachates, effects mammals, **151**:80
Rubber tire leachates, effects plants, **151**:80
Rubber tire leachates in aquatic environment, **151**:67 ff.
Rubber tire leachates, toxicity, **151**:80 ff.
Rubber tires, recycling, **151**:72
Rubber, vulcanization accelerators, **151**:70
Rumex acetosella (sorrel), atmospheric biomonitor, **157**:6
Runoff, rubber tire debris, **151**:107

S-glutathion-atrazine, soil sorption, **151**:132
Salinity, metal bioavailability effects marine mussels, **151**:40
Salmonella, spices irradiation control, **154**:19
Salmonella spp., poultry meat irradiation control, **154**:32
Salmonella typhimurium, Ames test for mutagenicity, **158**:114
Salmonella typhimurium, meat irradiation sensitivity, **154**:27, 38
Salvinia auriculata (floating plant), aquatic mercury biomonitor, **157**:34
Samarium, trace contamination estuaries, **155**:73 ff.
Sampling methods, plant atmospheric biomonitors, **157**:8, 15
San Francisco Bay, trace metal pollution sources, **155**:74
San Joaquin River (Calif), chlorpyrifos levels, **160**:51, 98
Sarin (GB), described, **156**:9
Sarin (GB), estimated reference dose, **156**:91, 146, 150
Sarin (GB), physical/chemical properties, **156**:13
Sarinase (phosphofluoridate fluorohydrolase), **156**:21

Scandium, bioaccumulation freshwater insect larvae, **158**:133

Scirpus cubensis (floating plant), aquatic mercury biomonitor, **157**:35

Seafood, irradiation pathogen control, **154**:37

Seafood, irradiation preservation, **154**:35

Seafood, shelf-life extension irradiation, **154**:36

Seals, pathological effects PCBs, **157**:108

Sediment contaminants, microbial assays, **159**:41 ff.

Sediment extraction methods, **159**:48

Sediment extracts, incubation microbial assay, **159**:49

Sediment pollutants, impact global element cycles, **159**:65

Sediment pollution assessment, microbial methods, **159**:41 ff.

Sediment sampling/storage, methods, **159**:46

Sediment toxicity assessment, microbial methods, **159**:45

Sediment-borne toxicity, chlorpyrifos, **160**:69, 76

Sediment-water interface, defined, **155**:124

Sediments, DDE residues Rio Grande Basin, **158**:35, 37

Sediments, elutriate microbial assay, **159**:48

Sediments, extraction methods, **159**:48

Sediments, mercury residues Rio Grande Basin, **158**:36, 38

Sediments, microbial methods contaminants, **159**:41 ff.

Sediments, polychlorinated alkanes contamination, **158**:77, 81

Sediments, pore waters microbial assay, **159**:48

Sediments, sampling & storage methods, **159**:46

Sediments, selenium residues Rio Grande Basin, **158**:36, 39

Sediments, trace metal levels vs freshwater insect larvae, **158**:134

Sediments, trace metal residues Rio Grande Basin, **158**:36

Selenastrum capricornutum, freshwater alga, **159**:99

Selenium, assimilation efficiency marine mussels, **151**:44

Selenium, contamination Rio Grande Basin biota, **158**:10, 16

Selenium, naturally occurring Rio Grande Basin, **158**:3

Selenium, residues birds Rio Grande Basin, **158**:16

Selenium, residues fish Rio Grande Basin, **158**:30

Selenium, residues invertebrates Rio Grande Basin, **158**:34

Selenium, residues plants Rio Grande Basin, **158**:35

Selenium, residues reptiles Rio Grande Basin, **158**:20

Selenium, residues Rio Grande water, **158**:41

Selenium, residues sediments Rio Grande Basin, **158**:36, 39

Selenium, Rio Grande Basin pollution, **158**:1 ff.

Selenium, trace contamination estuaries, **155**:73 ff.

Senescence delay, fresh fruit irradiation, **154**:2

Sensitive taxa, ecological role chlorpyrifos risk assessment, **160**:107

Sensory panel, irradiated food quality, **154**:28

Septicemia, opportunistic causal pathogens, **152**:72

Sequential extraction, trace metals estuarine sediment, **155**:90

Serratia marcescens, radiation sensitivity food, **154**:11

Seston loads, trace elements marine mussels, **151**:53

Sewage sludge, polychlorinated alkanes contamination, **158**:77, 81

Sharks, heavy metal levels Gulf of Mexico, **157**:78

Sheepshead minnow, rubber tire leachate toxicity, **151**:86, 89, 94

Shelf-life extension, food irradiation, **154**:2, 14

Shelf-life extension, meat irradiation, **154**:26
Shelf-life extension, seafood irradiation, **154**:36
Silver, assimilation efficiency marine mussels, **151**:44
Silver mining, environmental mercury contamination, **157**:28
Silversmiths, mercury exposure, **157**:44
Site reactivity probes, alkyl halide dehalogenation, **155**:49
Smoking, cadmium intake cigarettes, **154**:68
Sodium cyanide, drinking water effects in rats, **156**:119
Soil arthropods, general sensitivity insecticides, **154**:123
Soil bacteria, alkyl halide transformations, **155**:47
Soil biodegradation factors, atrazine, **151**:123
Soil desorption, atrazine, **151**:128
Soil fauna, field recovery pesticide application, **154**:111
Soil fauna, pesticide risk assessment, **154**:83 ff.
Soil fauna, pesticides predicted vs observed recovery, **154**:118
Soil leaching, ethalfluralin, **153**:72
Soil macropores, atrazine percolation, **151**:138
Soil, mercury contamination via air, **157**:31
Soil residues, ethalfluralin, **153**:82
Soil sorption, atrazine, **151**:125
Soil thin-layer chromatography, atrazine, **151**:132
Soil water content, atrazine binding, **151**:131
Soil-bound residues, ethalfluralin, **153**:70
Soil-bound residues, trifluralin, **153**:15
Soman (GD), estimated reference dose, **156**:100, 146, 150
Soman (GD), physical/chemical properties, **156**:14
Sorption/desorption, atrazine metabolites, **151**:132

South Korea, organochlorine pesticide residues food, **152**:26
Spanish moss, atmospheric mercury monitoring, **157**:33
Species variation, nerve agent effects, **156**:21
Species variation, RBC-ChE activity, **156**:22
Spices, irradiation bacterial/mold counts, **154**:20
Spices, irradiation plastic pouches, **154**:19
Spices, irradiation preservation, **154**:18
Spirillum volutans, motility inhibition test tire leachate, **151**:85
Sprout inhibition, vegetable irradiation, **154**:14
Staphylococcus aureus, D-Values food irradiation, **154**:7, 27, 39
Starch, irradiation breakdown to sugars, **154**:15
States of water, **155**:120
Stereochemistry, PHAs, **159**:7
Stoneflies (Perlodidae), trace metals vs sediment levels, **158**:134
Stoneflies, larval bioaccumulation heavy metals, **158**:132
Stoneflies, Pb, Zn, Cu vs sediment/water levels, **158**:137
Street runoff, tire leachate toxicity, **151**:106
Streptococcus faecalis, seafood irradiation control, **154**:39
Streptococcus faecium, radiation/temperature sensitivity, **154**:10
Streptococcus, radiation/temperature sensitivity, **154**:10
Stress proteins, sediment pollutant assays, **159**:66
Stressor characteristics, chlorpyrifos aquatic systems, **160**:12
Striped bass, *Marone saxitilis*, **159**:99
Striped bass, molinate toxicity, **159**:98
Striped bass, thiobencarb toxicity, **159**:98
Strontium, contamination Rio Grande Basin biota, **158**:10
Styrene polymers, automobile tire composition, **151**:68

Styrene-butadiene rubber, automobile tire composition, **151**:68
Styrene-butadiene rubber, chemistry, **151**:69
Styrene-butadiene rubber derivatives, carcinogenicity, **151**:82
Styrene-butadiene rubber tire dust, composition, **151**:82
Sub-particulate organic matter (SPOM), (illus.), **155**:118
Subchronic toxicity, CK, **156**:114
Subchronic toxicity, GA, **156**:68
Subchronic toxicity, GB, **156**:80
Subchronic toxicity, GD, **156**:95
Subchronic toxicity, HD, **156**:31
Subchronic toxicity, lewisite, **156**:103, 109
Subchronic toxicity, VX, **156**:55
Sublethal effects, pesticides soil fauna, **154**:83 ff.
Sublethal toxicity indicators, polychlorinated alkanes, **158**:112
Substitution, environmental dehalogenation processes, **155**:5
Sulfur mustard agents, physical/chemical properties, **156**:11
Sulfur mustard, environmental fate soil, **156**:131
Sulfur mustard H, described, **156**:9
Sulfur mustard HD, described, **156**:9
Sulfur mustard HT, described, **156**:9
Sulfur mustard, vapor effects eyes, **156**:33
Surface runoff, trifluralin, **153**:22, 24
Surface water runoff, atrazine, **151**:134
Survey methods, Rio Grande Basin biota, **158**:8
Survival curves, bacteria food irradiation, **154**:6
Synergism, chlorpyrifos with other agrochemicals, **160**:64
Synthetic rubber, chemistry, **151**:69

T sulfur mustard, acute toxicity, **156**:50
T sulfur mustard, chronic toxicity, **156**:50
T sulfur mustard, estimated reference dose, **156**:146
T sulfur mustard, genotoxicity, **156**:51
T sulfur mustard, physical/chemical properties, **156**:11
Tabun (GA), described, **156**:9
Tabun (GA), estimated reference dose, **156**:74, 146, 150
Tabun (GA), physical/chemical properties, **156**:13
Tar balls, crude oil marine sources, **153**:94
Taste panel, irradiated food quality, **154**:28
TCLP (Toxicity Characterization Leaching Procedure), USEPA, **151**:75
TCP (trichloro-2-pyridinol), structure, **160**:25
TCP, see trichloro-2-pyridinol, **160**:7
Temperature, metal bioavailability effects marine mussels, **151**:40
Teratogenic effects, cadmium, **154**:68
Teratogenicity, HD, **156**:46
Teratogenicity, polychlorinated alkanes, **158**:105
Termite control, chlorpyrifos environmental concentrations, **160**:104
Tesla (T), measure of magnetic field intensity, **159**:116
Test methods, pesticide risk assessment soil fauna, **154**:85
Texas, herbicide use, **158**:3
Thailand, DDT/HCH residues food, **152**:23
Thailand, PCB residues food, **152**:23
Thermal degradation, polychlorinated alkanes, **158**:70
Thiobarbituric acid, irradiated meat, **154**:27
Thiobencarb, rice herbicide fish kill California, **159**:97
Thiobencarb, striped bass toxicity, **159**:98
Thiol-Ca^{2+} peroxidation, mustard agent mechanism, **156**:17
Tiers (EPA), chlorpyrifos environmental concentrations, **160**:27
Tile drains, atrazine field leaching, **151**:144
Tillage effects, atrazine persistence, **151**:148
Tillandsia usneoides (Spanish moss), air mercury monitoring, **157**:33

Tin, trace contamination estuaries, **155:**73 ff.
Tire debris, atmospheric deposition/runoff, **151:**107
Tire dust, composition, **151:**81
Tire manufacturing, waste products, **151:**74
Tires, rubber leachates aquatic environment, **151:**67 ff.
Tires, rubber recycling, **151:**72
Tires, scrap landfills accumulation problems, **151:**70
Titanium oxide, rubber additive, **151:**69
Total DDT residues (DDT-R), **151:**6
Total suspended solid loads, metal bioavailability effects aquatic organisms, **151:**40
TOXI-Chromotest®, *Escherichia coli* tire leachate, **151:**85, 99
Toxicity assessment, sediments microbial methods, **159:**45
Toxicity Characterization Leaching Procedure (TCLP), USEPA, **151:**75
Toxicity mechanisms, CK (cyanogen chloride), **156:**25
Toxicity mechanisms, lewisite, **156:**25
Toxicity mechanisms, mustard agents, **156:**16
Toxicity mechanisms, nerve agents, **156:**18
Toxicity, polychlorinated alkanes, **158:**96, 98
Toxicity symptoms, CK (cyanogen chloride), **156:**25
Toxicity symptoms, lewisite, **156:**25
Toxicity symptoms, mustard agents, **156:**15
Toxicity symptoms, nerve agents, **156:**17
Toxicokinetics, polychlorinated alkanes, **158:**88
Trace element assimilation efficiency marine mussels, **151:**44, 52
Trace element concentrations, marine mussels model-predicted, **151:**52
Trace metal absorption efficiency marine mussels, **151:**44, 51
Trace metal adsorption, clay minerals, **155:**85
Trace metal adsorption, humic substances, **155:**86
Trace metal adsorption, hydrous metal oxides, **155:**86
Trace metal adsorption, organic matter, **155:**85
Trace metal adsorption processes, estuaries, **155:**85
Trace metal analysis, estuarine sediment samples, **155:**88
Trace metal bioaccumulation, freshwater insect larvae, **158:**129 ff.
Trace metal estuarine pollutants, listed, **155:**80
Trace metal estuarine pollution, chemical forms/levels, **155:**76
Trace metal flocculation, rivers estuaries, **155:**83
Trace metal levels, rivers/oceans, **155:**79
Trace metal levels, UK estuaries, **155:**77
Trace metal levels, US Atlantic estuaries, **155:**78
Trace metal levels, US estuaries, **155:**80
Trace metal pollution, principal sources estuaries, **155:**74
Trace metal pollution sources, San Francisco Bay, **155:**74
Trace metal-particle interactions, **155:**81
Trace metal-sediment estuarine dynamics, **155:**73 ff.
Trace metal-sediment estuarine pollutants, listed, **155:**80
Trace metal-sediment pollution assessment, **155:**73 ff.
Trace metals, anthropogenic sources estuaries, **155:**100
Trace metals, contamination Rio Grande Basin biota, **158:**10
Trace metals, estuarine bottom sediments, **155:**92
Trace metals, estuarine contamination case studies, **155:**97
Trace metals, natural sources estuaries, **155:**100
Trace metals, removal processes estuaries, **155:**96
Trace metals, residues invertebrates Rio Grande Basin, **158:**33

Trace metals, residues Rio Grande water, **158**:41
Trace metals, residues sediments Rio Grande Basin, **158**:36
Trace metals, sediment levels vs freshwater insect larvae, **158**:134
Trade names, trifluralin, **153**:5
Transgenic insect cells, PHA production, 159:15
Transgenic plants, PHA production, **159**:14
Transgenic production, PHAs, **159**:13
Trees, biomonitors atmospheric pollution, **157**:5
Triangle test, irradiated food quality, **154**:28
Triazine herbicides, interaction with chlorpyrifos, **160**:64
Trichinella spiralis, irradiation destruction, **154**:2
Trichinella spiralis, pork meat irradiation control, **154**:22, 25
Trichinosis, pork meat irradiation control, **154**:22, 25
Trichloro-2-pyridinol (TCP), chlorpyrifos metabolite, **160**:7
Trichloro-2-pyridinol (TCP), structure, **160**:25
Trifluralin, adsorption/desorption, **153**:19 ff.
Trifluralin, aerobic soil transformation, **153**:13
Trifluralin, air/dust concentrations, **153**:49
Trifluralin, air/rain/lake concentrations, **153**:48
Trifluralin, airborne residues growing season, **153**:40
Trifluralin, analytical sensitivities environmental samples, **153**:62
Trifluralin, anerobic soil transformation, **153**:14
Trifluralin, annual field soil carryover percentages, **153**:33
Trifluralin, benzimidazole formation, **153**:10
Trifluralin, biphasic dissipation from soils, **153**:37
Trifluralin, CAS number, **153**:5
Trifluralin, chemical names, **153**:5

Trifluralin, chemical structure, **153**:88
Trifluralin, crops annual usage, **153**:4
Trifluralin, dealkylation steps, **153**:14
Trifluralin, dissipation in water/sediments, **153**:17
Trifluralin, dissipation pathway chart, **153**:8, 10
Trifluralin, environmental dissipation air/rainfall, **153**:39
Trifluralin, environmental dissipation soil, **153**:33
Trifluralin, environmental dissipation water/sediments, **153**:38
Trifluralin, environmental exposure assessment, **153**:42
Trifluralin, environmental fate, **153**:1 ff.
Trifluralin, groundwater contamination potential, **153**:23
Trifluralin, half-lives air, **153**:11
Trifluralin, half-lives soil, **153**:35
Trifluralin, half-lives vapor state, **153**:41
Trifluralin, hydrolysis, **153**:5
Trifluralin, $K_{oc}s$ chart, **153**:19
Trifluralin, $K_{ow}s$, **153**:5, 20, 22
Trifluralin, leaching soil, **153**:20
Trifluralin, major manfacturers, **153**:2
Trifluralin, metabolite stability, **153**:15
Trifluralin metabolites, chemical nomenclature, **153**:63
Trifluralin, microorganism degradation, **153**:13
Trifluralin, mobility soil, **153**:19 ff.
Trifluralin, mode of action, **153**:2
Trifluralin, monitoring water sources, **153**:45
Trifluralin, photodecomposition soil, **153**:6, 8
Trifluralin, photodecomposition water, **153**:7, 9
Trifluralin, physicochemical properties, **153**:5, 88
Trifluralin, principal crops used, **153**:2
Trifluralin, rainfall/irrigation runoff effects, **153**:24
Trifluralin, reduction steps, **153**:14
Trifluralin, residues soil, **153**:43
Trifluralin, residues vegetation, 42
Trifluralin, residues water/sediments/surface films, **153**:44

Trifluralin, river basin drainage loads, **153**:26
Trifluralin, river basin sampling, **153**:47
Trifluralin, sediment degradation, **153**:16
Trifluralin, snow-borne, **153**:40
Trifluralin, soil biological/chemical transformations, **153**:12
Trifluralin, soil diffusion, **153**:28
Trifluralin, soil transformation pathway chart, **153**:8
Trifluralin, soil-bound residues, **153**:15
Trifluralin, surface runoff, **153**:22, 24
Trifluralin, surface runoff treated watersheds, **153**:27
Trifluralin, trade names, **153**:5
Trifluralin, transformation processes, **153**:5
Trifluralin, trifluoromethyl oxidation, **153**:14
Trifluralin, use patterns, **153**:3
Trifluralin, vapor concentrations over treated fields, **153**:49
Trifluralin, vapor-phase photolysis, **153**:12
Trifluralin, volatility losses from soil, **153**:28
Trifluralin, water photolysis pathway chart, **153**:10
Trifluralin, watershed residues Manitoba, Canada, **153**:27
Trifluralin, weeds controlled, **153**:2
Trifluralin, well water concentrations U.S., **153**:46
Trifolium repens (clover), atmospheric biomonitor, **157**:6
Triphenyltin, no-observed-effect concentrations soil fauna, **154**:108
Triphenyltin, risk assessment soil fauna, **154**:91
orse bacterium, *Agrobacterium tumefaciens*, **159**:14
Tsetse fly control, organochlorine insecticides Africa, **151**:2

U.S.-Mexico border, population, **158**:6
UK estuaries, trace metal levels, **155**:77
Uncertainty Factor, oral reference dose derivation, **156**:28
Uruguay, human lead exposure, **159**:25 ff.
Uruguay, lead contamination, **159**:25 ff.
Uruguay, map, **159**:27
US Atlantic estuaries, trace metal levels, **155**:78
US estuaries, trace metal levels, **155**:80

Vanadium, trace contamination estuaries, **155**:73 ff.
Vapor-phase photolysis, ethalfluralin, **153**:68
Vapor-phase photolysis, trifluralin, **153**:12
Variovorax paradoxus, PHA biodegradation, **159**:17
Vegetable crops, cadmium contamination, **154**:55 ff., 62
Vegetable crops, cadmium levels, **154**:64
Vegetables, cadmium levels vs soil pH, **154**:62
Vegetables/fruits, irradiation preservation, **154**:12
Vegetables, high cadmium affinity, **154**:62
Vegetables, sprouting inhibition irradiation, **154**:14
Vegetation burning, major mercury source, **157**:29
Very high current configuration (VHCC), child cancer, **159**:114
Very low current configuration (VLCC), child cancer, **159**:117
VHCC, see Very high current configuration, **159**:114
Vibrio cholera, irradiation sensitivity seafood, **154**:37
Vibrio fisheri, luminescent marine bacterium, **159**:56
Vibrio parahaemolyticus, irradiation sensitivity seafood, **154**:38
Vibrio vulnificus, irradiation sensitivity seafood, **154**:38
Victoria amazonica (floating plant), aquatic mercury biomonitor, **157**:35
Vietnam, organochlorine pesticide residues food, **152**:25
Vitamin B_{12s}, (illus.), **155**:37
VLCC, see Very low current configuration, **159**:117

Volatility losses, ethalfluralin field studies, **153**:73
Volatility losses, trifluralin field studies, **153**:30
Volatilization, atrazine field, **151**:151
Volatilization, atrazine soil surface, **151**:133
Volcanoes, haloorganics syntheses, **155**:3,4
Vulcanization accelerators, rubber tire manufacture, **151**:70
Vulcanization, rubber tire chemistry, **151**:69
VX (nerve agent), physical/chemical properties, **156**:13
VX, acute toxicity, **156**:54
VX, bird toxicity, **156**:142
VX, carcinogenicity, **156**:61
VX, developmental/reproductive effects, **156**:60
VX, estimated reference dose, **156**:62, 146, 149
VX, fish toxicity, **156**:141
VX, genotoxicity, **156**:61
VX, hydrolysis, **156**:134
VX, hydrolysis soil, **156**:139
VX, hydrolytic pathway (diagram), **156**:135
VX, hydrolytic products, **156**:135
VX, neurotoxicity, **156**:59
VX, photolysis, **156**:134
VX, RBC-ChE activity fed sheep, **156**:57
VX, RBC-ChE activity injected rats, **156**:56
VX, RfD *vs* human toxicity data comparison, **156**:67
VX, subchronic toxicity, **156**:55
VX warfare agent, described, **156**:9

Water, bound, natural organic matter, **155**:115 ff.
Water column, defined, **155**:123
Water contamination, polychlorinated alkanes, **158**:76, 78
Water, drinking heavy metals permissible levels US, **157**:85
Water hyacinth (*Eichhornia crassipes*), mercury biomonitor, **157**:34
Water, states of, **155**:120

Water surface biofilms, defined, **155**:122
Well water contamination, trifluralin, **153**:46
Wheat, DDT/HCH residues India, **152**:3
Wheat, insect disinfestation irradiation, **154**:3
White sturgeon, *Acipenser transmontanus*, **159**:99
Wine, cadmium content, **154**:66
Wire codes, EMF child brain tumors, **159**:118
Wire codes, HCC/LCC child brain tumors comparison, **159**:119
Wire configuration code, power lines distance from homes, **159**:114
Wood preservatives, PCDE source, **157**:133
Worker lead exposure, **159**:30

Xanthomonas maltophilia, as opportunistic pathogen, **152**:62
Xanthomonas maltophilia, occurrence drinking water, **152**:62
Xanthomonas maltophilia, oral infective dose humans, **152**:62
Xanthomonas maltophilia, pneumonia causal pathogen, **152**:71
Xanthomonas, opportunistic pathogens drinking water, **152**:58

Yersinia enterocolitica, red meat irradiation control, **154**:22
Ytterbium, trace contamination estuaries, **155**:73 ff.

Zinc, assimilation efficiency marine mussels, **151**:44
Zinc, bioaccumulation freshwater insect larvae, **158**:131
Zinc, contamination Rio Grande Basin biota, **158**:10
Zinc, EDTA environmental removal, **152**:102
Zinc, organism levels Gulf of Mexico, **157**:77
Zinc oxide, rubber additive, **151**:69
Zinc, pollution Gulf of Mexico, **157**:65, 73
Zinc sulfate, rubber additive, **151**:70